W9-BSB-198

UNDERSTANDING SCIENCE & NATURE

# Weather & Climate

TIME-LIFE
ALEXANDRIA, VIRGINIA

C O N T E N T S

# 1 The Air Above



# 2 The Air in Motion



# 3 Storm Machines

# 1

# The Air Above

Although people take the air for granted, Earth's atmosphere is the planet's most precious treasure. Born shortly after the planet formed, some 4.6 billion years ago, the atmosphere was created as gases escaped from the planetesimals that collided to create Earth. About 3 billion years ago, plants began to photosynthesize, altering the primitive atmosphere by releasing vast quantities of oxygen.

Over time, the atmosphere developed a complex structure consisting of several different layers extending to an altitude of about 600 miles. Weather occurs in the bottom layer, called the troposphere. In the troposphere, air pressure and temperature decline as altitude increases. Above, in the stratosphere, temperature rises with increasing altitude. About 15 miles above the surface, a thin layer of ozone gas absorbs some of the harmful ultraviolet light emitted by the sun, protecting organisms below. Above all of it is the ionosphere, a region of ionized gases that reflect electrical waves from the planet's surface, making possible radio communications.

Just as the atmosphere once changed in response to plant life, it is now changing as a result of human activities. Pollution from cars and factories, and even from aerosol sprays, has threatened the delicate balance of gases. The inhabitants of Earth have begun to realize that the atmosphere is as fragile as it is precious.

**The rising sun** is reddened and distorted by Earth's atmosphere. The atmosphere filters and softens the sun's radiation, making life possible.

# How Did the Atmosphere Form?

Today, Earth's atmosphere is a mixture of gases—78 percent nitrogen, 21 percent oxygen, and small amounts of other gases such as carbon dioxide. When the planet was first formed, however, it had no oxygen, and its gases were those that were present in the early Solar System.

Earth was born when small, rocky bodies known as planetesimals—formed from the dust and gas of the solar nebula—smashed into each other and gradually took shape as a planet. As the planet grew, the gases trapped within the planetesimals escaped, enveloping the globe. Over time, the first plants began to release oxygen and the primitive atmosphere evolved into today's thick blanket of air.

**Billions of years ago,** thick mats of primitive algae released oxygen into the atmosphere. They are preserved today as fossils called stromatolites.

**The birth of the atmosphere**

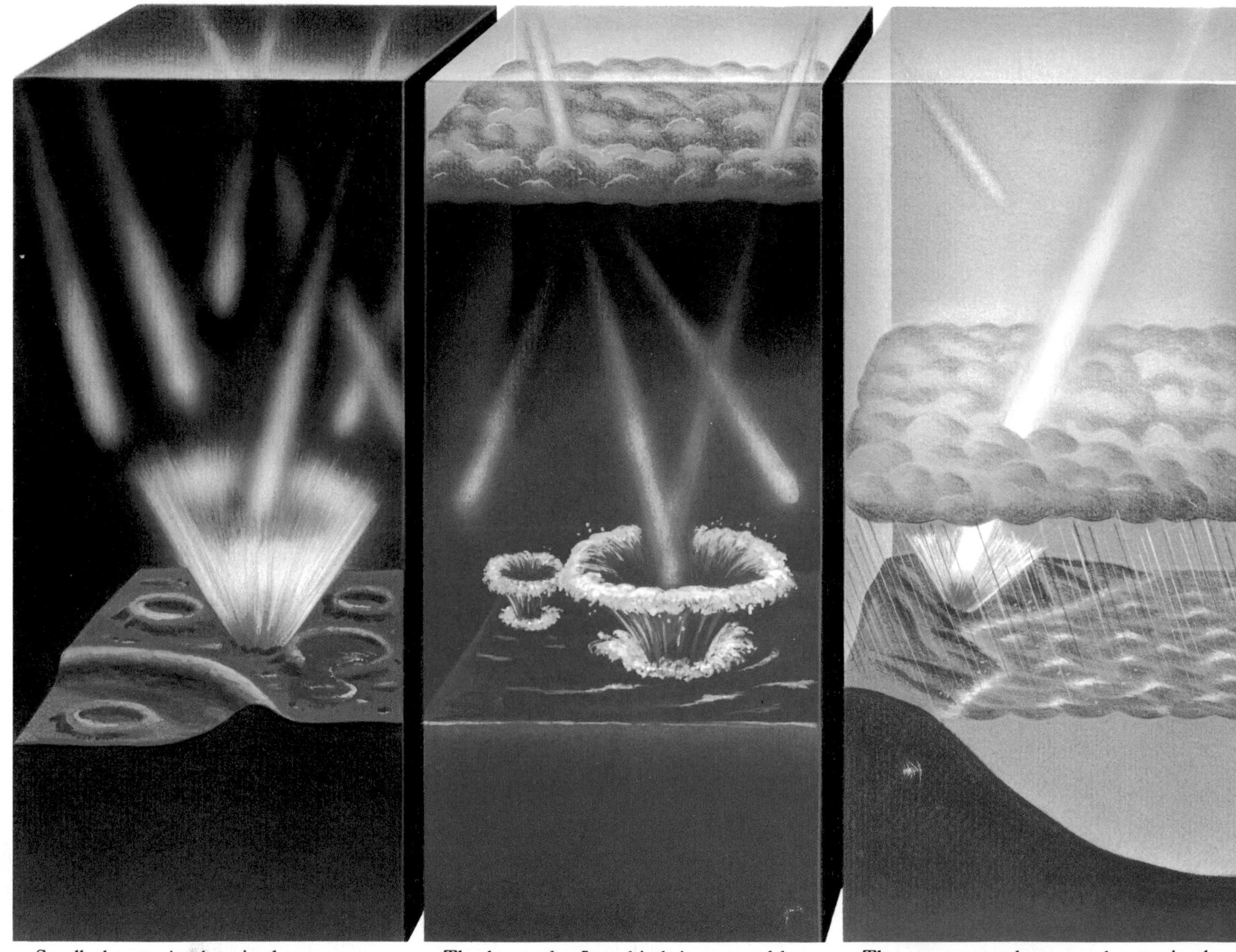

Small planetesimals rain down upon the infant Earth of 4.6 billion years ago. Gases of the solar nebula, trapped within the planet, are released upon impact to form Earth's original atmosphere of nitrogen, carbon dioxide, and water vapor.

The heat of a fiery birth is trapped by the thick clouds of the primitive atmosphere. The "greenhouse gases," such as carbon dioxide and water vapor, prevent heat from radiating away into space. The Earth's surface melts in a fiery sea of molten magma.

The oceans are born as planetesimal impacts become less frequent and the Earth begins to cool. Water vapor condenses from the thick clouds, and rains that continue for eons gradually fill the lowlands and give birth to the first seas.

## A volcanic genesis

According to one theory, volcanic activity dominated the surface of the young Earth. The early atmosphere may have formed as gases trapped within the planet's silicon mantle escaped through volcanic vents.

An early, airless Earth.

Volcanoes spew out gases.

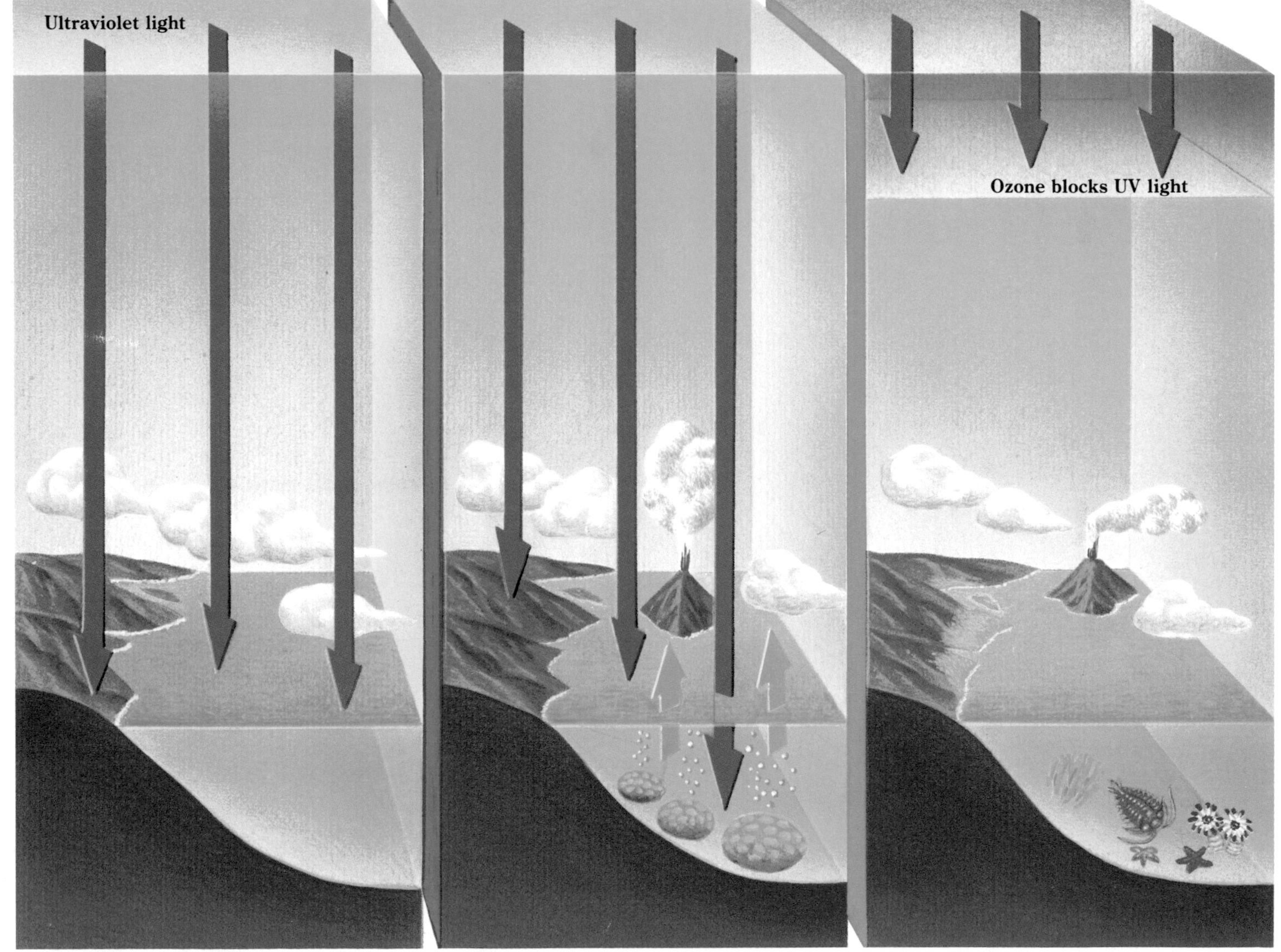


The air begins to clear as water vapor condenses to form the oceans. Over time, carbon dioxide dissolves into the oceans, leaving an atmosphere dominated by nitrogen. With no oxygen to form a protective ozone layer, ultraviolet light from the sun reaches the Earth's surface unimpeded.

Life appears in the primitive oceans within the first billion years. The simple blue-green algae are protected from ultraviolet rays by the seawater. The algae use sunlight and carbon dioxide to produce energy, releasing oxygen as waste. Slowly, oxygen begins to collect in the atmosphere.

Over billions of years, an oxygen-rich atmosphere evolves. Photochemical reactions in the upper atmosphere create a thin layer of ozone, which filters out harmful ultraviolet rays. Now life can move out of the oceans and onto land, where evolution creates a variety of complex organisms.

# How Is the Atmosphere Divided?

The air that shrouds the Earth has no visible boundaries. After studying it with instruments aboard rockets and balloons, however, scientists have realized that it can be divided into five distinct layers. Starting from the Earth's surface, these layers are the troposphere, stratosphere, mesosphere, thermosphere, and exosphere. As altitude increases, the density of the atmosphere declines, until, several hundred miles above the surface, it fades away.

**The exosphere,** beginning about 300 miles above the Earth's surface, is extremely thin, consisting mainly of light molecules of hydrogen and helium. There is no clear-cut border between the exosphere and outer space. Temperatures measure around 999° Kelvin (a scientific scale in which zero equals −459° F.). Because the air is so thin in the upper layers, however, even very high temperatures have little effect on spacecraft.

**The thermosphere,** 50 to 300 miles above the surface, contains just one-billionth of the total atmosphere. The ionosphere, divided by scientists into lettered regions, begins here and descends into the mesosphere.

**The mesosphere,** 30 to 50 miles above the surface, is the coldest region of the atmosphere. It includes the lowest part of the ionosphere. The composition of the mesosphere is basically the same as that of the lower layers.

**The stratosphere** is a very stable region about 10 to 30 miles above the surface. With almost no water vapor present, few clouds form here. At an altitude of about 15 miles, the ozone layer absorbs most of the sun's ultraviolet rays. The temperature increases with altitude in the stratosphere.

**The troposphere** varies in height according to season and latitude, reaching a maximum of about 10 miles at the equator. About 80 percent of the total atmospheric mass is found here. Temperature decreases with increasing altitude.

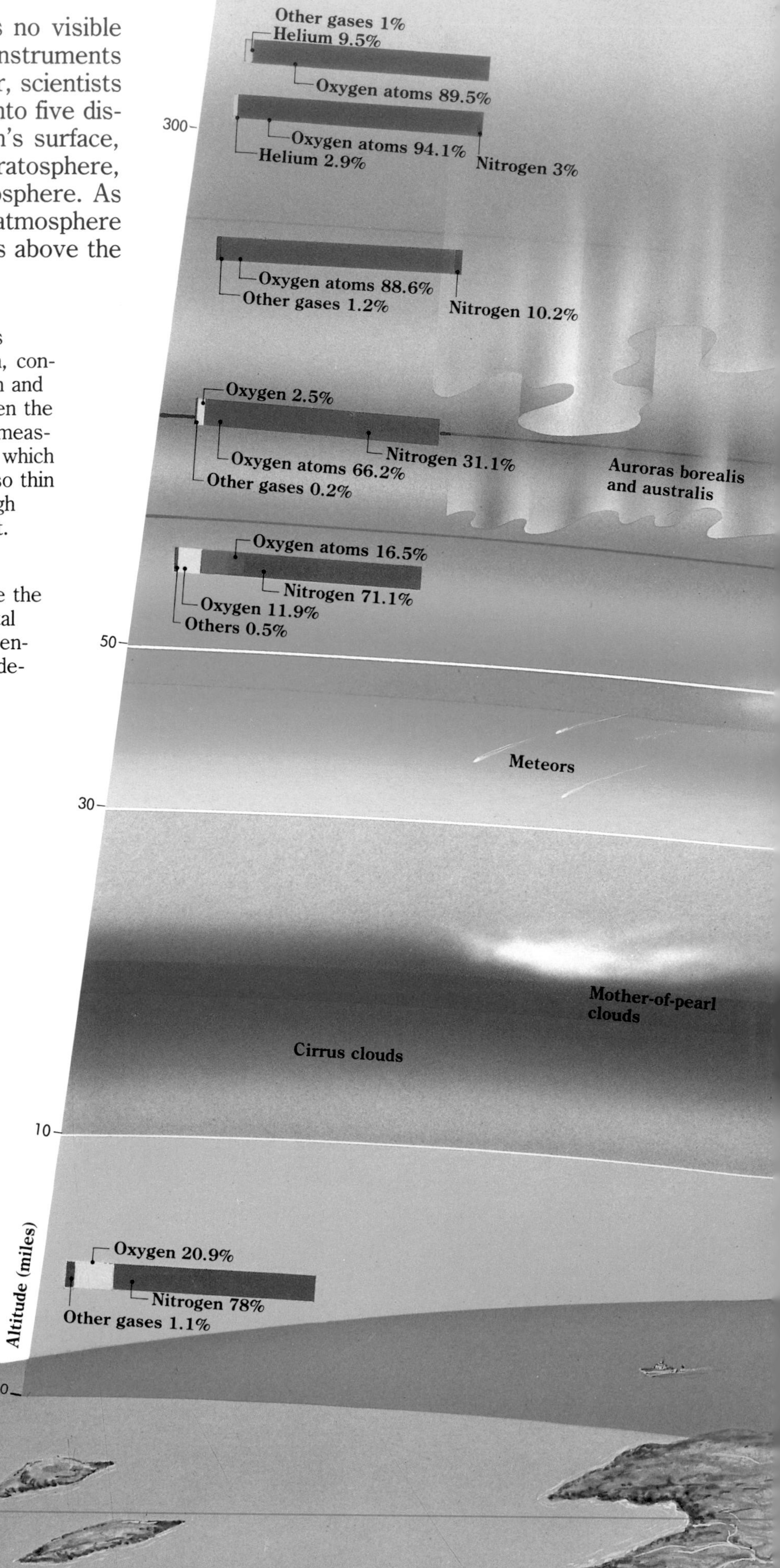

Temperature
(Kelvin)
999.24K
995.83K
F$_2$ layer
976.01K
F$_1$ layer
854.56K
E layer
195.08K
Noctilucent clouds
D layer
198.64K
270.65K
Ozone layer
221.55K
Cumulonimbus
clouds
216.65K
288.15K (15°C)
Cosmic rays
X-rays
Ultraviolet
Visible light
Infrared
Radio
Exosphere
Thermosphere
Mesosphere
Stratosphere
Troposphere

# Where Does the Atmosphere End?

Seen from space, the atmosphere is an ethereal veil, barely bound to the Earth through the force of gravity. From the planet's surface, though, the air seems impossibly high, fading off to deep blue at the edge of space. In actual fact, the atmosphere is several hundred miles deep, and it has no definite edge.

The atmosphere is densest in the lowest layer, the troposphere, and becomes progressively thinner as altitude increases. Between the Earth's surface and the beginning of the thermosphere, at an altitude of about 50 miles, the composition of the air remains constant, with nitrogen and oxygen making up 99 percent of the gases. The thermosphere contains most of the ionosphere, where incoming solar radiation ionizes the atmospheric gas—that is, the atoms and molecules lose one or more electrons and become electrically charged. At higher levels, solar x-rays and ultraviolet rays break up large molecules, and nitrogen and oxygen become increasingly scarce. Above about 300 miles, in the exosphere, nothing remains but atoms of hydrogen and helium. The exosphere marks the end of the atmosphere—beyond it lies the magnetosphere, a vast, airless region dominated by the force of the Earth's magnetic field.

**Seen from space,** the atmosphere, several hundred miles deep, appears to be a thin blue mist shrouding the planet.

Magnetosphere

Plasma

Van Allen belts

▼**A U.S. polar orbit** weather satellite collects data on temperature, water vapor, and clouds from 525 miles up.

Thermosphere

Mesosphere

Cumulonimbus clouds

Stratosphere

Troposphere

**Charged particles**—called plasma—in the ionosphere rise into the upper atmosphere during the day. At night, with no solar energy, the plasma sinks back to the ionosphere.

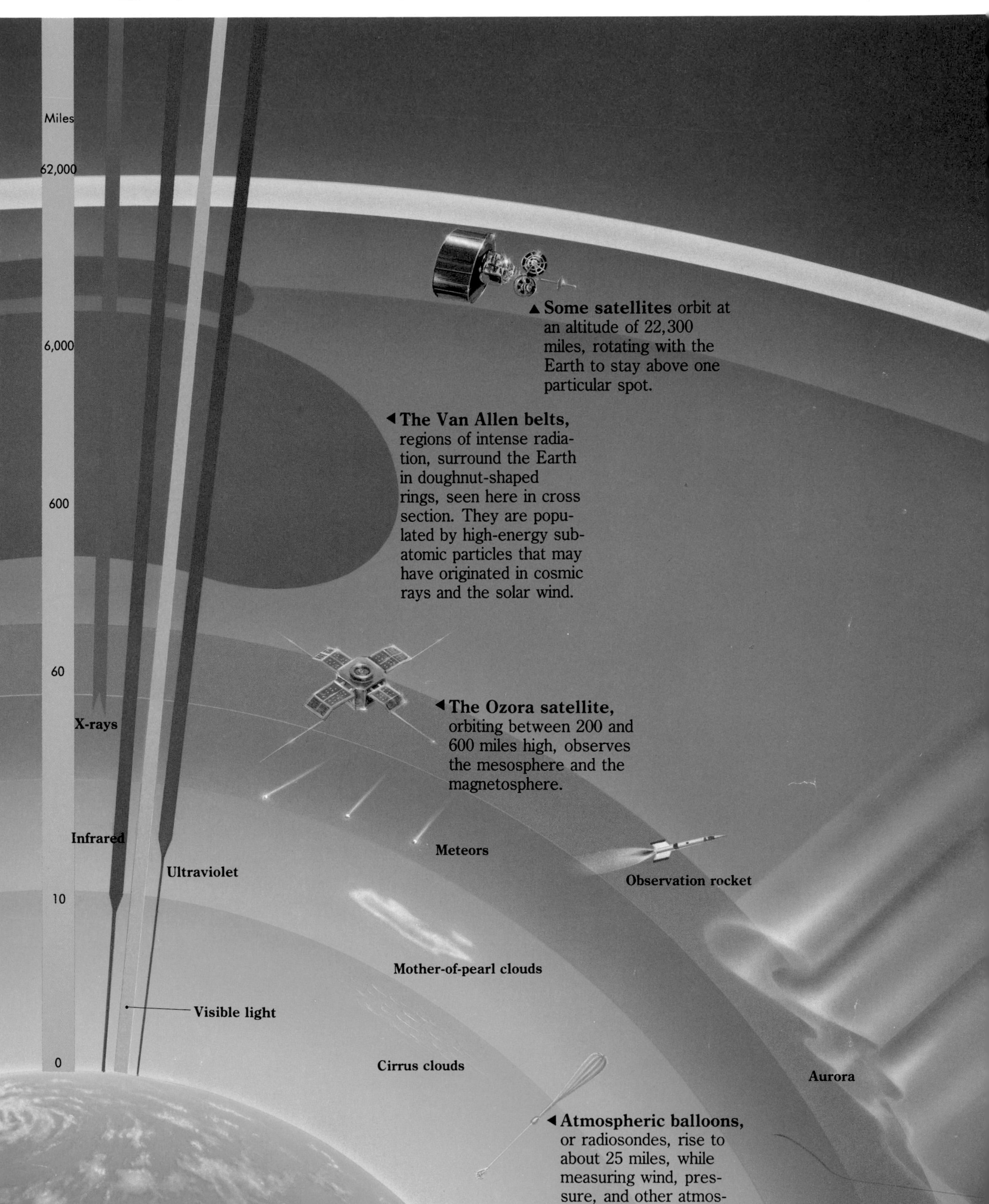


▲ **Some satellites** orbit at an altitude of 22,300 miles, rotating with the Earth to stay above one particular spot.

◀ **The Van Allen belts,** regions of intense radiation, surround the Earth in doughnut-shaped rings, seen here in cross section. They are populated by high-energy subatomic particles that may have originated in cosmic rays and the solar wind.

◀ **The Ozora satellite,** orbiting between 200 and 600 miles high, observes the mesosphere and the magnetosphere.

◀ **Atmospheric balloons,** or radiosondes, rise to about 25 miles, while measuring wind, pressure, and other atmospheric phenomena.

# What Is the Ozone Layer?

The sun and stars produce many kinds of radiation that are harmful to living things—including ultraviolet light, an invisible form of radiation that falls between violet light and x-rays in the electromagnetic spectrum. Fortunately, the Earth's atmosphere blocks most damaging radiation from the planet's surface. Its ozone layer, particularly, plays a vital role in absorbing the short ultraviolet wavelengths produced by the sun.

At an altitude that varies from as low as 6 miles to as high as 35 miles up, the ozone layer is a thin band of the atmosphere in which solar ultraviolet light reacts with oxygen molecules to create ozone gas. The gas represents less than one-millionth of the total atmospheric volume: The entire ozone layer would be just a tenth of an inch thick if it were subjected to the atmospheric pressure at the surface of the Earth. Yet without this minor component of the atmosphere, humans would be afflicted with more instances of skin cancers and blindness, and crops would wither as the radiation broke down organic molecules. Indeed, without ozone, life itself would probably be impossible on Earth.

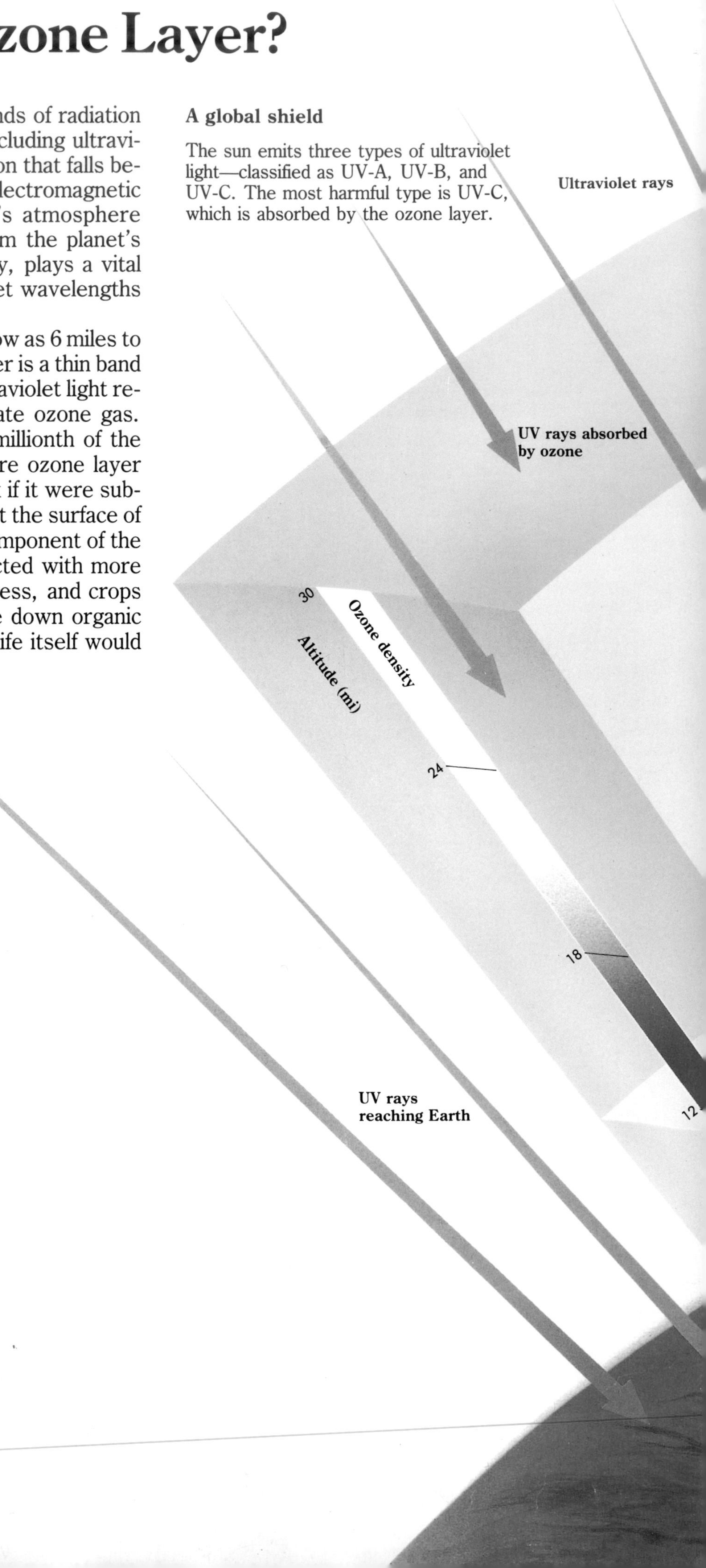

**A global shield**

The sun emits three types of ultraviolet light—classified as UV-A, UV-B, and UV-C. The most harmful type is UV-C, which is absorbed by the ozone layer.

Ultraviolet rays

UV rays absorbed by ozone

30

Ozone density

Altitude (mi)

24

18

12

UV rays reaching Earth

**An unprotected world**

Ultraviolet light can destroy genetic material in living cells, causing cancers and mutations. If the Earth had no ozone layer, its surface might resemble that of the planet Mars, where there is no ozone and—scientists believe—no life.

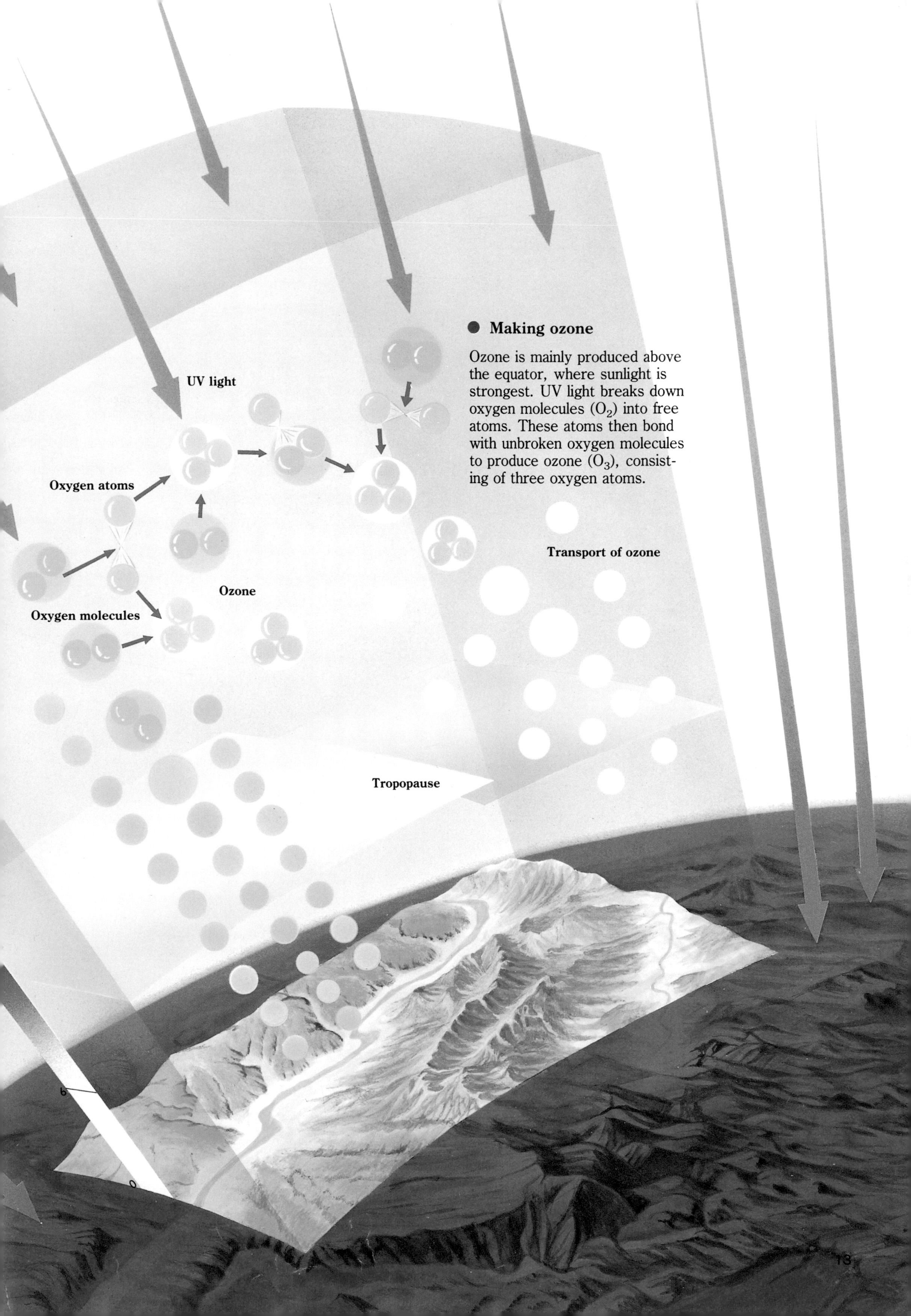

UV light

Oxygen atoms

Ozone

Oxygen molecules

## Making ozone

Ozone is mainly produced above the equator, where sunlight is strongest. UV light breaks down oxygen molecules ($O_2$) into free atoms. These atoms then bond with unbroken oxygen molecules to produce ozone ($O_3$), consisting of three oxygen atoms.

Transport of ozone

Tropopause

6

0

# What Is the Ionosphere?

In a region about 40 to 600 miles above the surface of the Earth, gas molecules in the atmosphere collide with high-energy particles from the sun. These collisions strip electrons from the molecules in a process called ionization. The result is a rapidly moving plasma consisting of electrically charged particles called ions.

Ionization occurs in the ionosphere, which consists of three layers, labeled D (the lowest), E, and F (the highest). These layers are subdivided according to the concentration of plasma within them. The lowest concentration is found in the $F_1$ layer, which is located at an altitude between 90 and 140 miles. The highest concentration of plasma is in the $F_2$ layer, which is at an altitude of about 140 to 300 miles. Because solar energy is required to create the ions, at night the $F_1$ layer disappears.

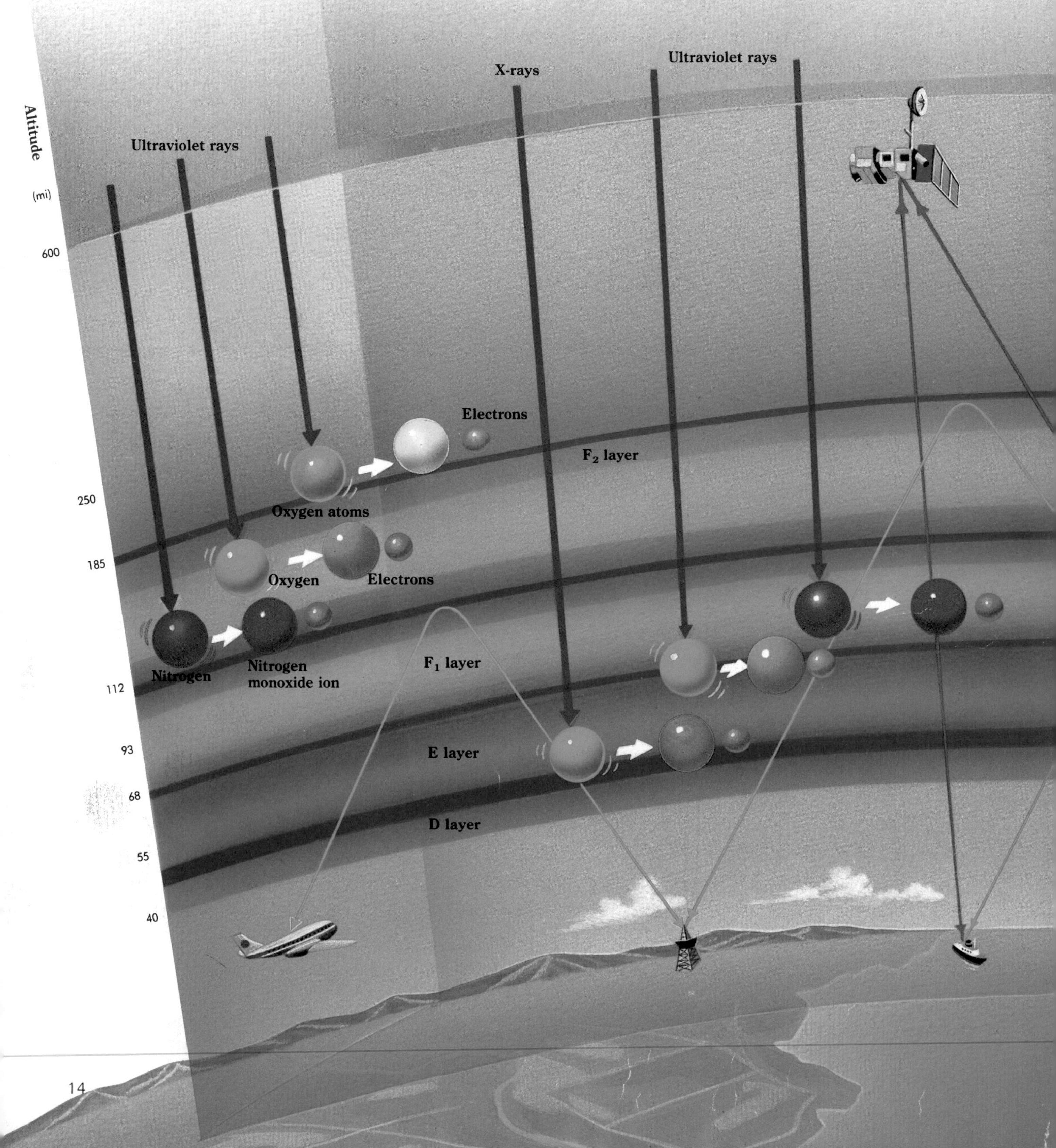

## The ionosphere and radio

The ionosphere makes globe-girdling shortwave radio transmissions possible. Short radio wavelengths are reflected off the F layer of the ionosphere, bouncing back to Earth thousands of miles from where they started. Longer wavelengths, which are used in commercial broadcasts, bounce off lower layers of the ionosphere and thus have a shorter range.

## The changeable ionosphere

The type and density of ions and electrons in the ionosphere vary according to altitude *(chart)* and time of day. During the day, the E layer of the ionosphere is clearly defined; at night it becomes indistinct. Part of the F layer disappears entirely during the night, and the ion population in general declines at night.

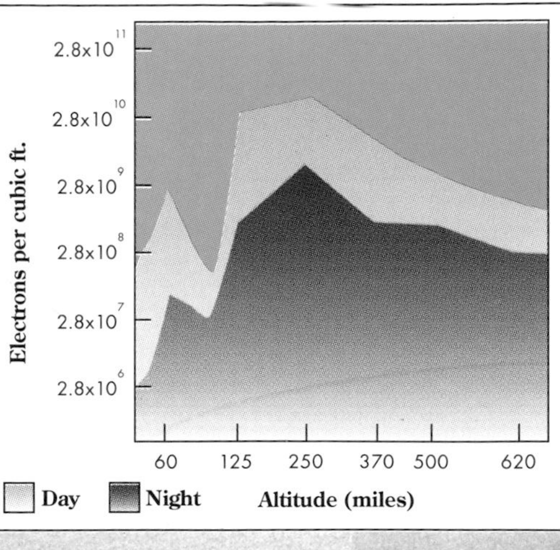


### Inside the ionosphere

In the E and D layers, x-rays and ultraviolet radiation strip oxygen and nitrogen molecules of electrons. The chemical reactions that result create ions of oxygen ($O_2+$) and nitrogen monoxide (NO+). $O_2+$ and NO+ ions are also present in the F layer, although the $F_2$ layer contains mostly oxygen ions.

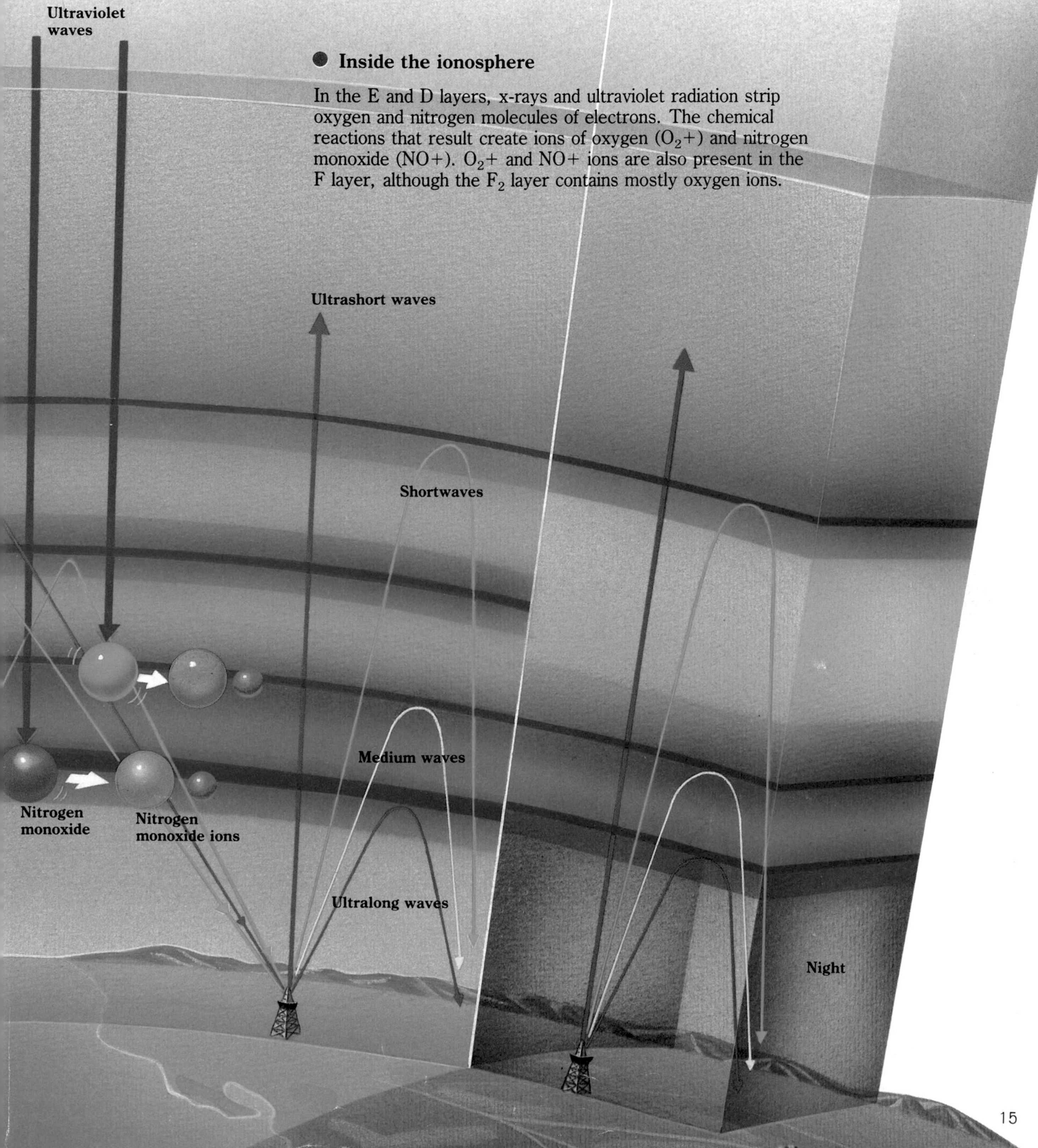

# What Causes the Aurora?

In the high latitudes of both the Northern and Southern hemispheres, the skies sometimes light up in spectacular displays known as the aurora borealis and aurora australis. These auroras are produced when electrically charged particles from the sun collide with atoms and molecules in the ionosphere at an altitude of 50 to 300 miles, causing them to emit light.

These solar protons and electrons travel to Earth on the solar wind, a powerful blast of particles that blows at one million miles an hour from the sun. The solar wind also carries part of the sun's magnetic field, which interacts with Earth's magnetosphere to allow the particles to plunge earthward along lines of magnetic force near the poles.

**A colorful curtain** of an aurora lights up the Alaskan sky.

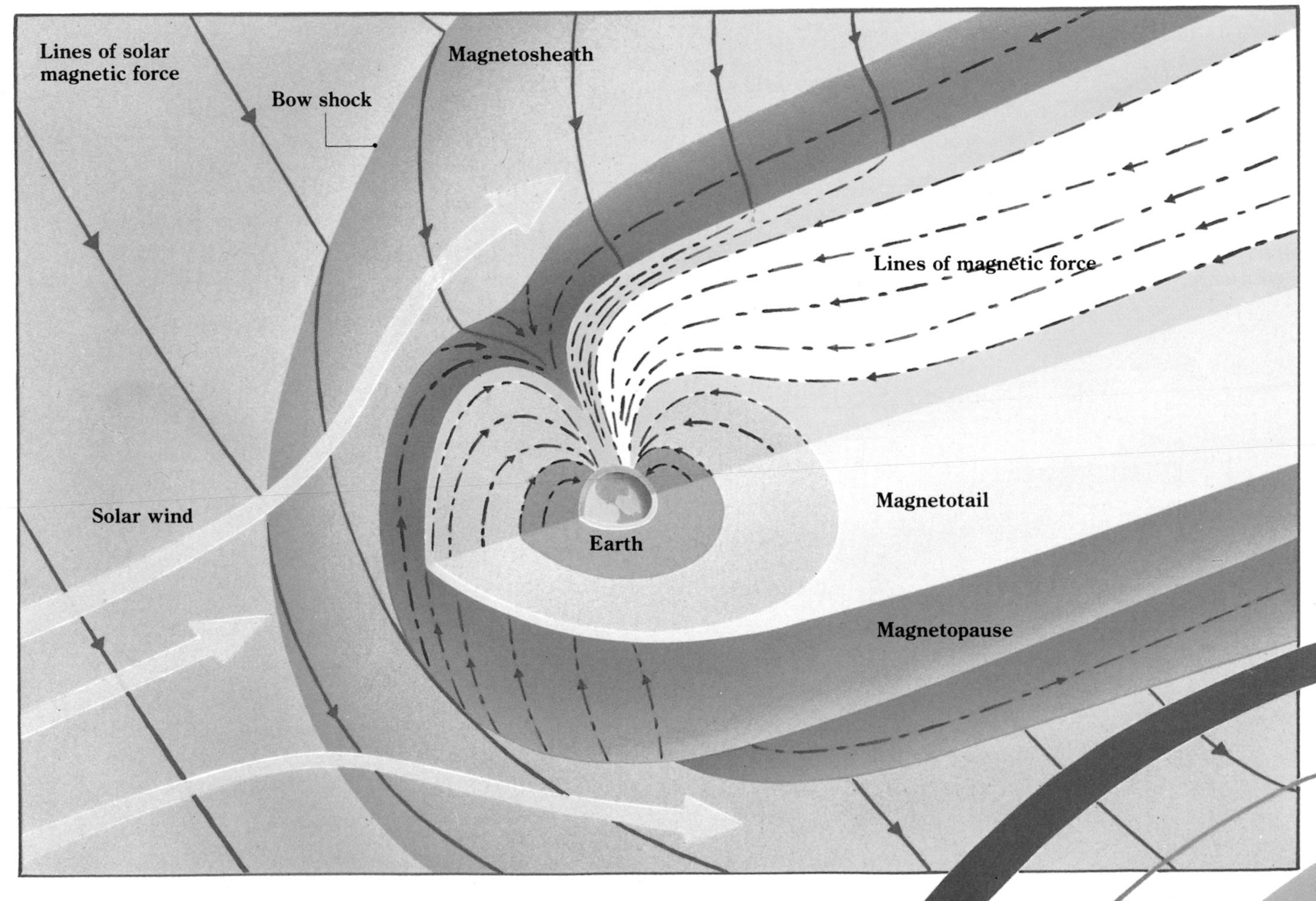


### A mingling of magnetic fields

Pressure from the solar wind compresses Earth's magnetic field on the daytime side to a distance of about 40,000 miles from the surface. On the nighttime side, the solar wind extends the magnetosphere a hundred times farther. The solar magnetic field connects with Earth's magnetic field at the boundary of the magnetosphere on the night side, and solar wind particles follow lines of magnetic force toward the poles.

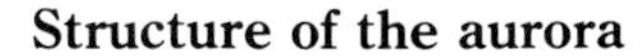

### The aurora generator

Protons in the solar wind have a positive electrical charge, electrons a negative charge. The different charges cause them to flow in opposite directions—protons toward the morning side of the magnetosphere, electrons toward the evening side. The particle flow creates oppositely charged poles, setting up what is, in effect, a huge electrical generator. The current flows through the auroral ovals in the magnetosphere above the planet's poles, in what is known as a field-aligned current. The total amount of electricity created in the "aurora generator" is more than a trillion watts.

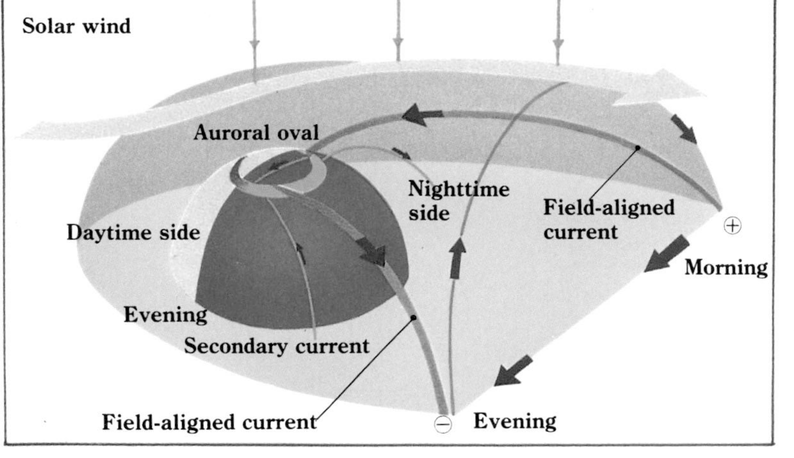


### Structure of the aurora

Electrons in the field-aligned current spiral around Earth's lines of magnetic force as they enter the atmosphere. There, they collide with atoms and molecules, exciting them and causing them to emit light. The incoming electrons are slowed by the collision and emit x-rays, while electrons from the excited gases release still more electrons in a chain reaction of collisions. Oxygen atoms excited in this manner emit green light, while nitrogen molecules produce a rosy pink light.

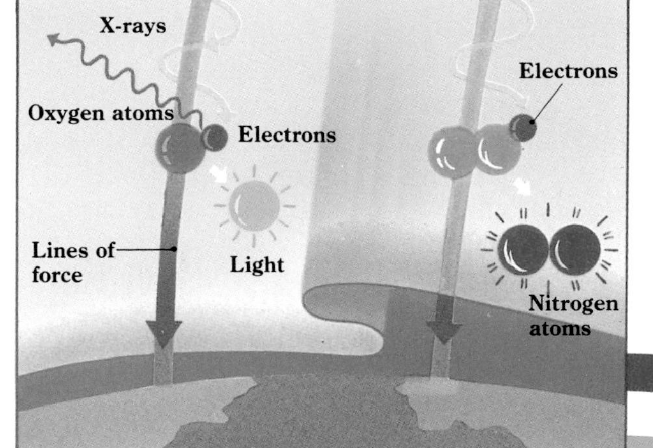


### Auroral currents

As seen from satellites orbiting far above the Earth, the auroras appear as oval-shaped bands surrounding the polar regions. In addition to the field-aligned current flowing on the inside of the oval, a secondary current flows in the opposite direction along the outside of the oval.

Auroral oval

Daytime side

Nighttime side

Morning

Evening

# What Are the Van Allen Belts?

America's first space satellite, *Explorer 1,* quickly proved the scientific value of space exploration when it was launched in 1958. Six hundred miles above the Earth's surface, an experiment aboard the spacecraft detected a belt of radiation 100 million times more intense than the natural background radiation on the ground. A second belt of radiation was later found at an altitude of 12,000 miles.

Together, these radioactive zones are known as the Van Allen belts, after physicist James Van Allen, whose experiment detected them. Made of charged particles from cosmic rays and the solar wind that are trapped in the Earth's lines of magnetic force, each one forms a torus (doughnut shape) around the Earth. The inner and outer belts differ according to the composition and energy levels of the particles trapped in them.

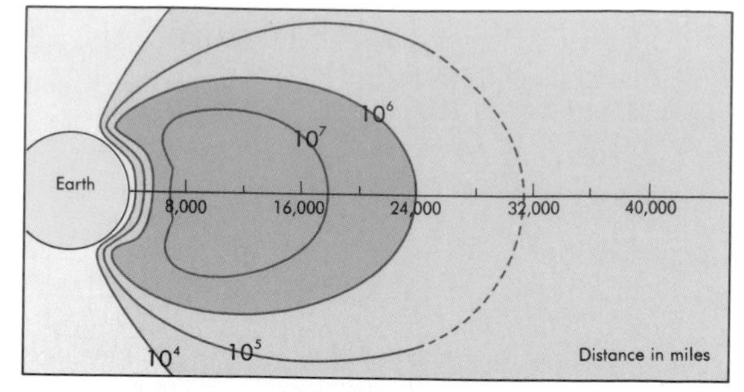


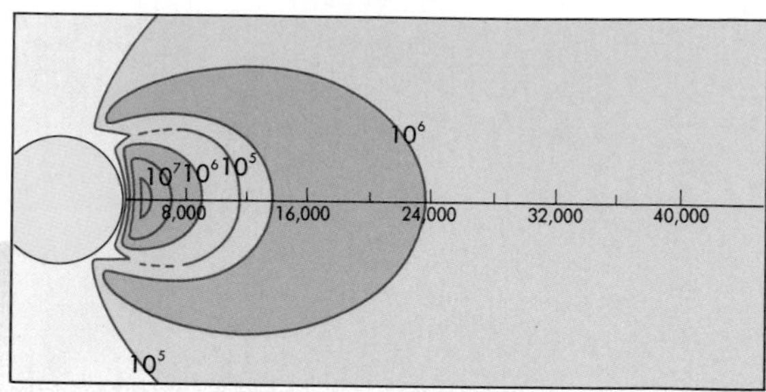


**High-energy protons** populate the Van Allen belts as shown in the upper diagram. The lower diagram shows high-energy electrons. (Darker regions have higher concentrations.)

### ● The inner Van Allen belt

Most of the protons and electrons that inhabit the inner Van Allen belt are created by the decay of neutrons. These, in turn, are the product of collisions between cosmic rays and the hydrogen and helium atoms in the upper atmosphere. The charged particles created in this way are caught in the Earth's magnetic field and possess a high-speed motion known as drift motion.

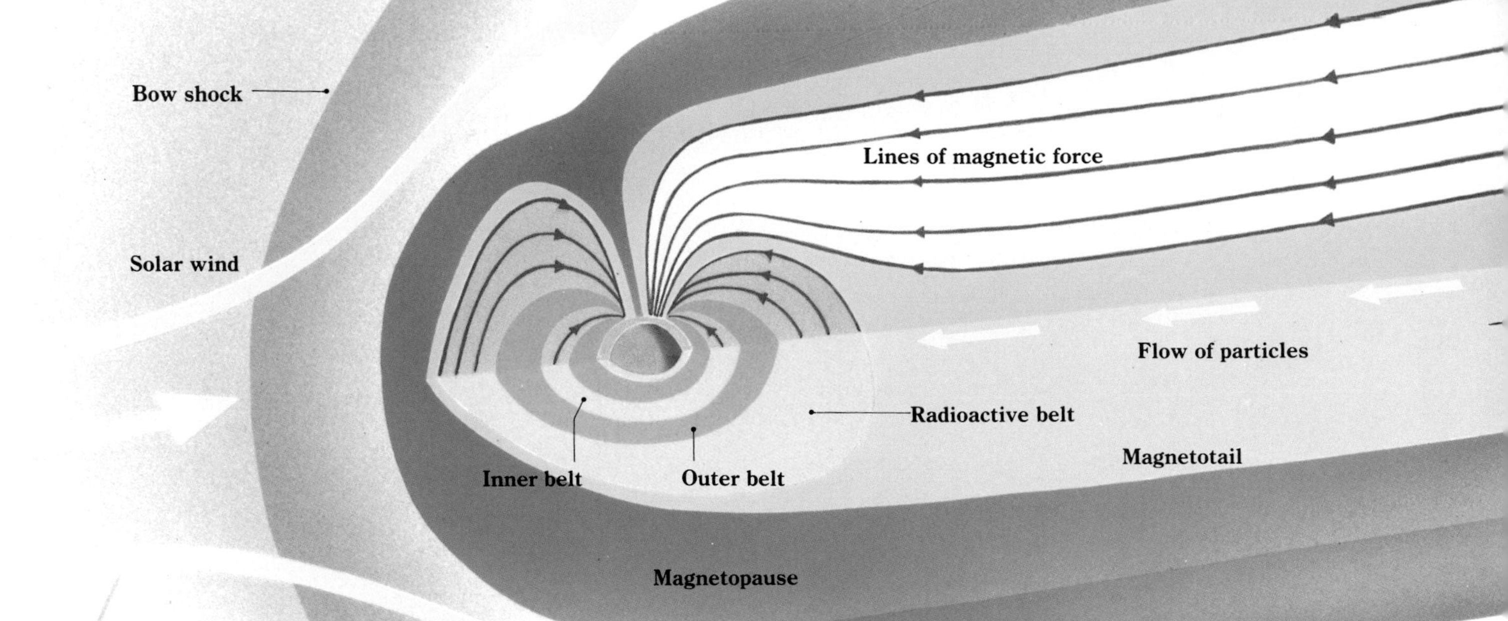


## Earth's magnetic field

Earth is like a huge magnet, and its lines of force form the magnetosphere. Pressure from plasma in the solar wind compresses the magnetosphere on the day side of the planet and stretches it out on the night side. The protons and electrons in the plasma are trapped in the magnetic field because of their electrical charge. The Van Allen belts extend from about 600 to 15,000 miles from Earth.

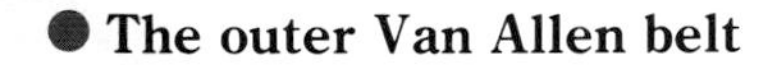


## The outer Van Allen belt

The inner and outer belts have slightly different compositions and energy levels. The outer belt accumulates most of its particles from the solar wind, rather than from cosmic rays. The inner Van Allen belt has fewer but more energetic protons than the outer belt.

## Structure of the belts

There is no sharp dividing line between the two Van Allen belts; the inner belt gradually merges with the outer belt. The two regions are distinguished by their differing populations of particles and their energy levels.

# Why Doesn't the Sun Burn the Earth?

In the unimaginably hot, crushingly dense core of the sun, 5 million tons of matter is converted into energy each second. Only a tiny fraction of that energy reaches the Earth; yet that amounts to 40 trillion calories every second.

If there were no atmosphere, this incoming energy would heat the Earth to 180° F. at the equator. Fortunately, clouds, the atmosphere, and the Earth's surface reflect about 34 percent of the solar energy back into space. Another 19 percent is absorbed by the clouds and atmosphere. Only 47 percent makes it all the way to the surface. Much of this energy is consumed by the evaporation of water to create clouds, which, in turn, keep the Earth cool.

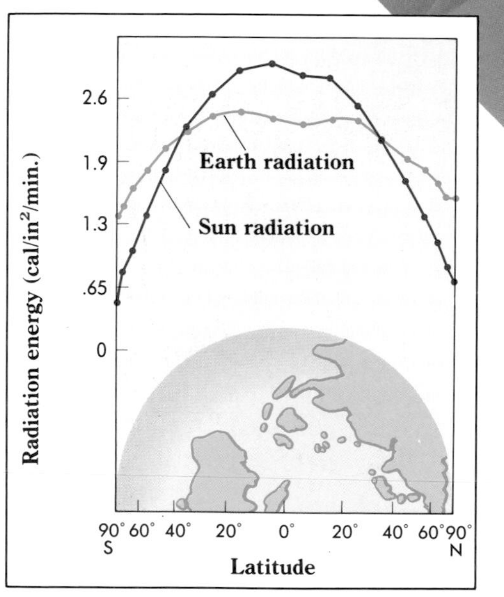


**The Earth is not heated** evenly by the sun. As shown above, near the equator the Earth receives more solar radiation *(pink)* than it loses. At the poles, the planet loses more energy than it receives *(blue)*.

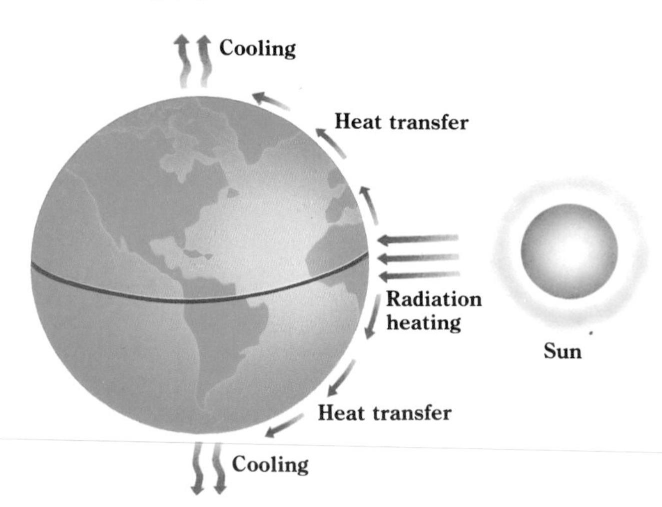


**Heat circulation**

The moving atmosphere and swirling ocean currents transport heat away from the equator, preventing overheating at those latitudes.

## A complex cycle

Less than half of the solar energy reaching Earth makes it to the surface. Some of that energy is transferred back into space. The rest is absorbed by the land, the ocean, and atmospheric gases such as water vapor and carbon dioxide. This pattern of energy transport *(shown at left in percentages)* keeps the Earth from becoming too hot or too cold.

# Why Is the Sky Blue during the Day but Red at Sunset?

Even though sunlight is white (meaning that it contains all colors together), the sky on a sunny day looks blue. This is because sunlight entering the atmosphere bumps into air molecules and dust particles, which cause different wavelengths of light to split off in a process called scattering. The sky looks blue on a clear day because small atmospheric particles scatter short blue wavelengths more than the long red wavelengths. However, at sunrise or sunset, and especially when the air is dusty, the sky looks red. This is because the sun's light must travel through the atmosphere for a greater distance when it is near the horizon. The blue light is bent away from the eye altogether, while large dust particles scatter red light to create a beautiful evening sky.

**Red dawns and brilliant sunsets** are caused by the way the atmosphere scatters long red wavelengths of light.

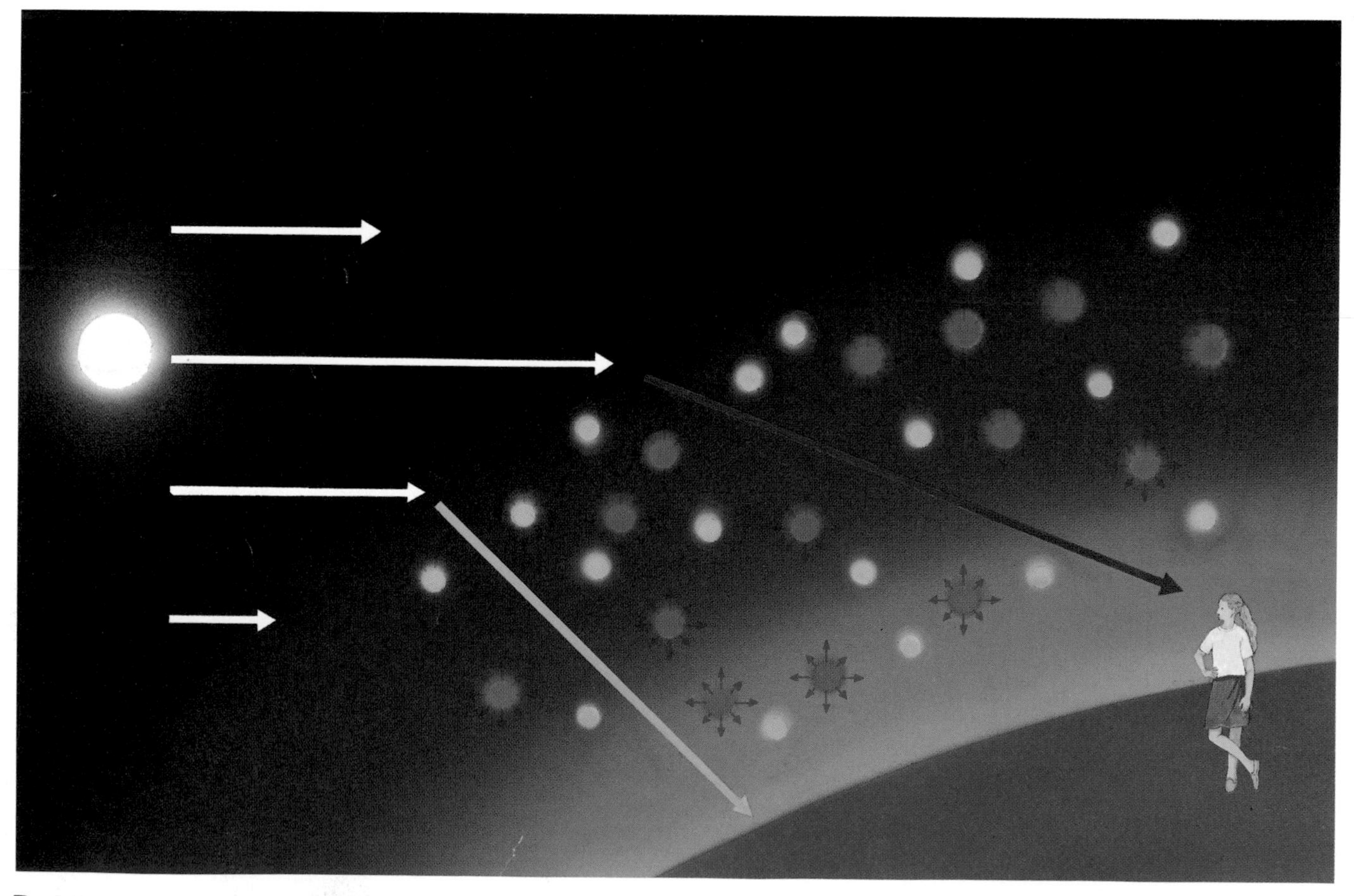

**Red light goes the distance**

When the sun is near the horizon, its light travels a long path through the atmosphere. Blue light is refracted away from the eye. Red light is refracted less and scatters off the large dust particles in the air.

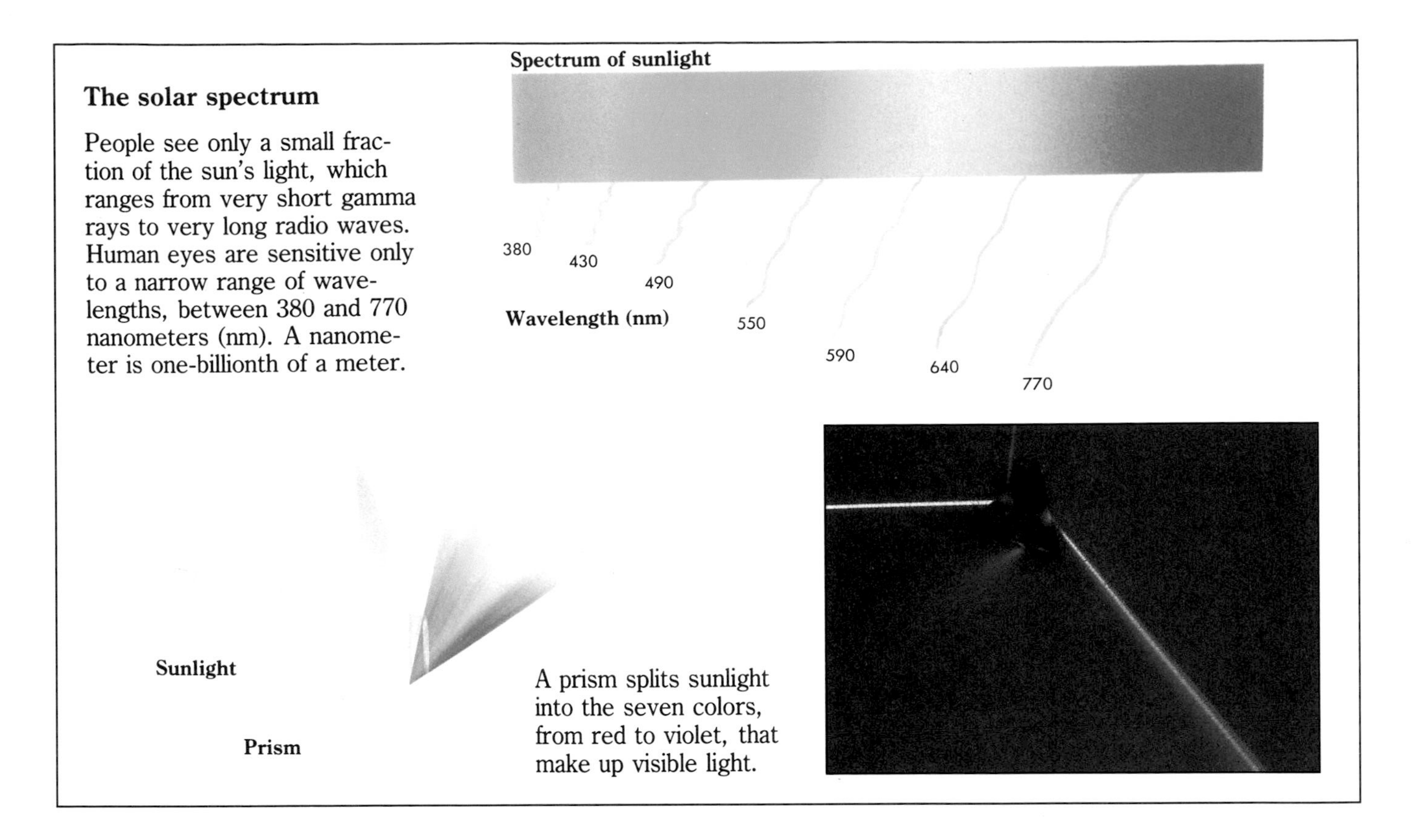


## The solar spectrum

People see only a small fraction of the sun's light, which ranges from very short gamma rays to very long radio waves. Human eyes are sensitive only to a narrow range of wavelengths, between 380 and 770 nanometers (nm). A nanometer is one-billionth of a meter.

A prism splits sunlight into the seven colors, from red to violet, that make up visible light.

## The birth of the blues

When the sun is overhead, its light has to pass through a comparatively thin layer of atmosphere. In a process called Rayleigh scattering, the small air molecules are much more effective at splitting off blue light than red. Blue light reaches the eye in greater volume than red light, so the sky looks blue.

# Why Is It Colder on Mountaintops?

The atmosphere is densest at low elevations. At higher elevations, the atmosphere is thinner and air pressure is lower. Rising air that has been heated by radiation from the surface expands and cools as it ascends into the region of lower air pressure, through a process known as adiabatic cooling. Typically, air cools by about 5.5° F. for every 1,000 feet of altitude. On a mountaintop 6,500 feet high, the air is about 35° F. cooler than at sea level. When clouds form in a rising air mass, the rate of cooling decreases to about one-half of that for a cloudless air mass. Taking the average for the entire troposphere, temperature decreases by about 3.5° F. for every 1,000 feet of altitude.

**Air pressure** is two-thirds of that at sea level atop 12,000-foot Mt. Fuji.

### Higher means colder

Solar radiation passes through the atmosphere to heat the surface of the Earth. Infrared radiation from the surface then warms the atmosphere above it. In the figure below, atmospheric layer C is warmed by heat from the surface. Layer C then radiates its heat both upward and downward. Above, layer B is warmed by heat released from layer C. In turn, layer B radiates its heat upward and downward, warming the atmosphere above and below it. But heat is lost

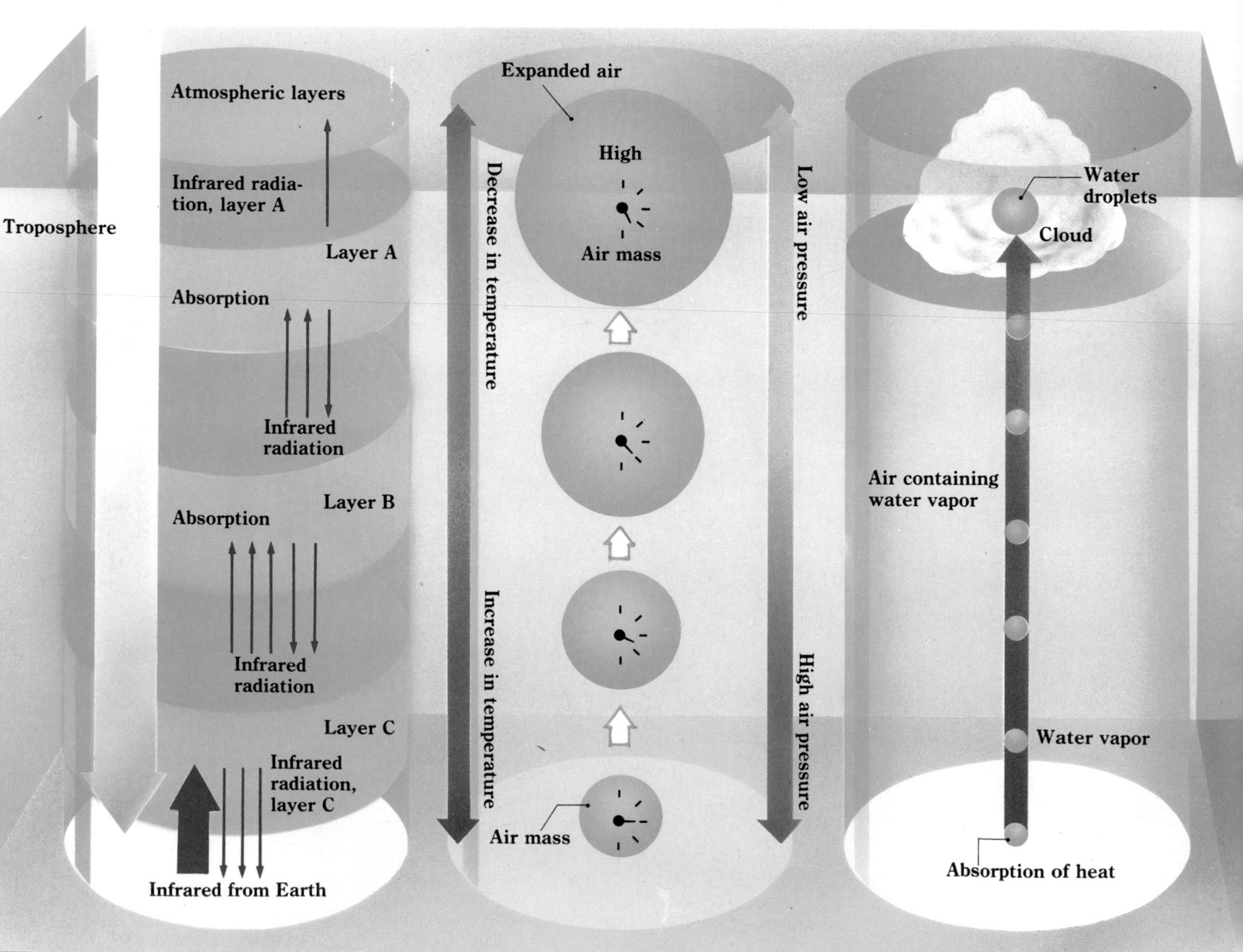

### A warming trend

The chart at right shows temperature changes from the surface to an altitude of about 30 miles. At the upper boundary of the troposphere, known as the tropopause, about 7 miles high, the temperature stops declining and begins to rise. In the stratosphere, additional heat is provided by the ozone layer. The ozone absorbs harmful ultraviolet radiation from the sun. In the process, energy generated by the ultraviolet rays heats the stratosphere.

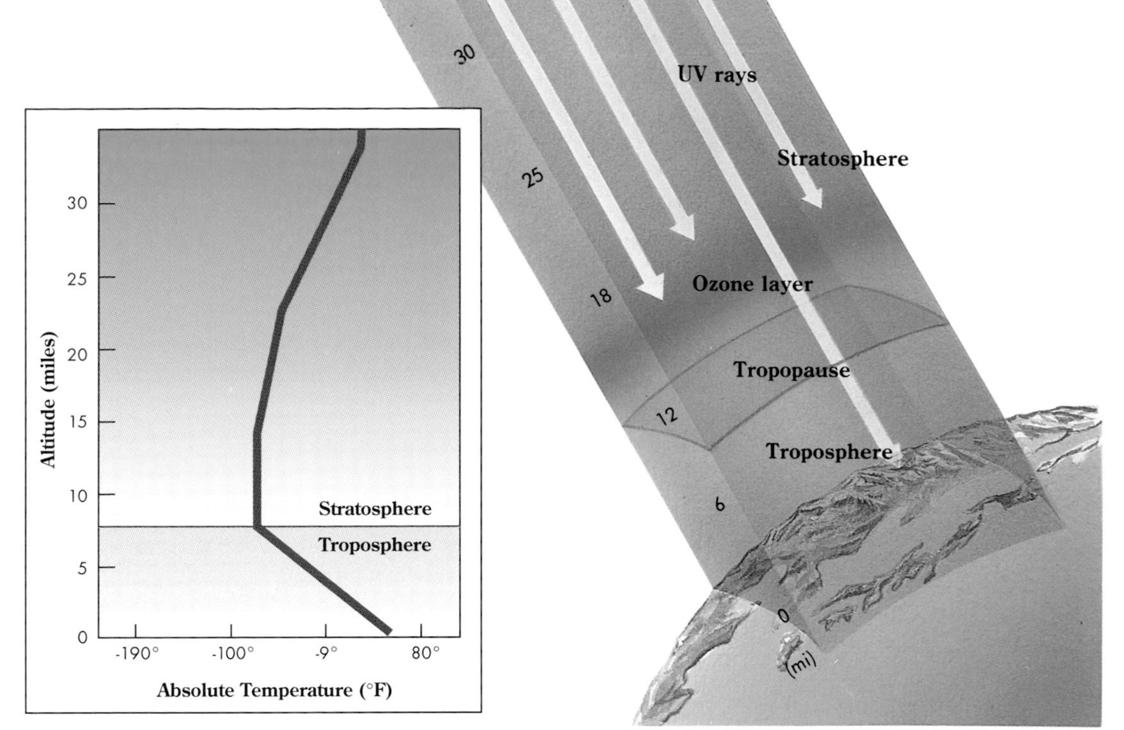


during the process, so the upper atmosphere receives a smaller share of the surface heat. Because the atmosphere is less dense at higher altitudes, rising air also cools as it expands through the mechanism called adiabatic cooling.

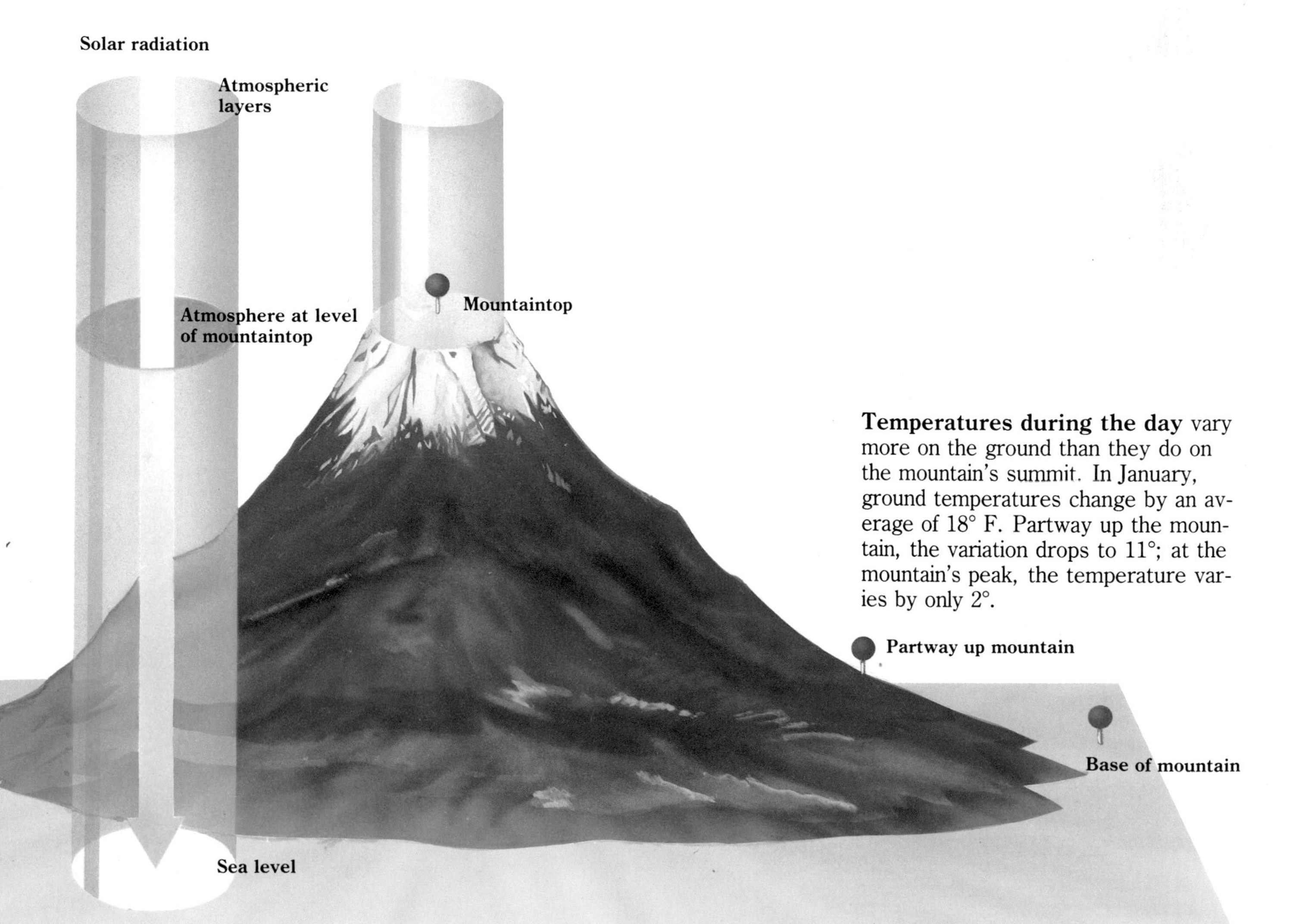


**Temperatures during the day** vary more on the ground than they do on the mountain's summit. In January, ground temperatures change by an average of 18° F. Partway up the mountain, the variation drops to 11°; at the mountain's peak, the temperature varies by only 2°.

# How Polluted Is the Atmosphere?

The atmosphere is a dirty place. It is filled with virtually everything that is light enough to be carried by the wind. Much of this pollution happens naturally as a result of dust storms, forest fires, and volcanic eruptions. But in recent years, human beings have dramatically increased the atmosphere's burden of pollutants. As populations around the world have grown, waste products from industry and agriculture have poured into the skies. Engines and furnaces burning oil, coal, and natural gas—the so-called fossil fuels—release a wide variety of pollutants. Chemical compounds, such as chlorofluorocarbons in refrigerators and aerosol sprays, not only pollute but actually destroy the atmosphere's ozone layer.

**A catalog of pollutants**

**Sulfur oxides** are compounds of sulfur and oxygen generated by fossil fuels.
**Nitrogen oxides,** produced by fossil fuel combustion, create photochemical smogs.
**Hydrogen chloride** is a volatile, strong-smelling gas used in some industrial processes.
**Dust particles** created by many processes float in the air and can cause respiratory illness when they are inhaled.
**Smoke and soot** contain carbon and tars, and toxic substances such as cadmium and lead.
**Aerosols** are tiny particles floating in suspension and may include many harmful substances.

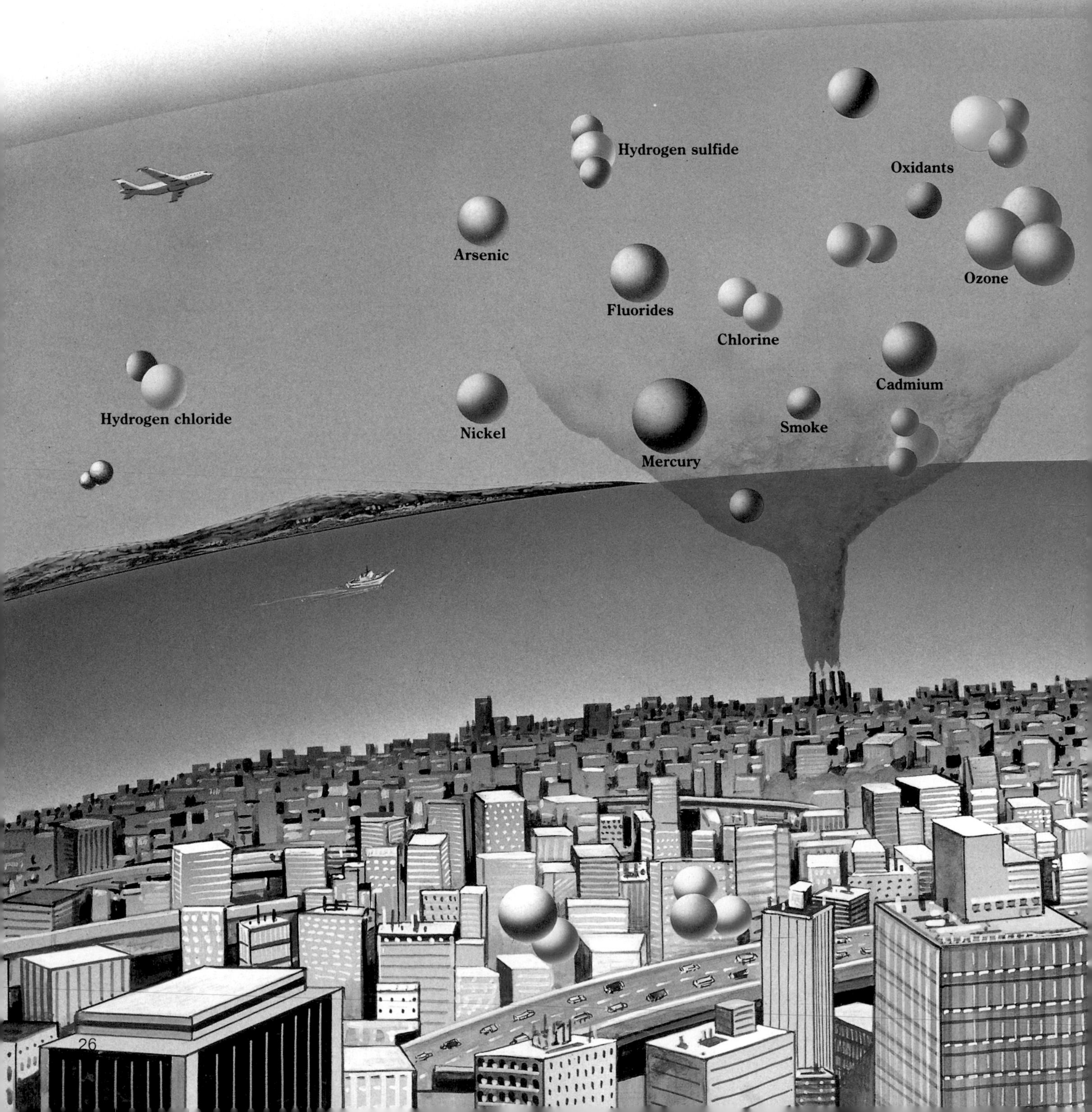

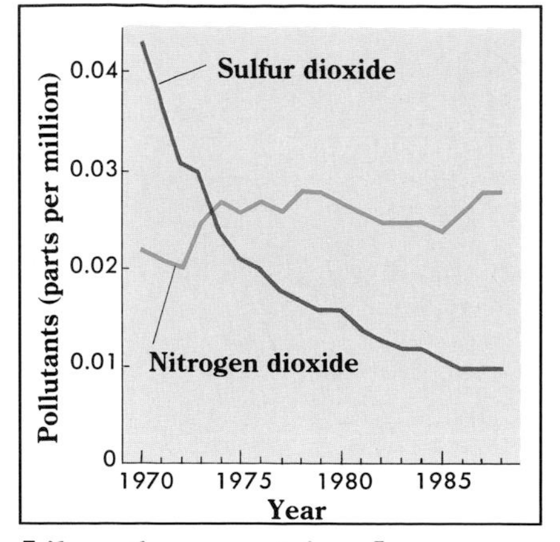


**Like other countries,** Japan has managed to reduce some kinds of air pollution, but not others, as shown in the chart above.

## Urban pollution

The graph at right shows levels of air pollution in some major urban areas from 1980 to 1985. Figures express the weight (in micrograms) of pollutants in 1 cubic meter of air. As population increases, so does the use of fossil fuels, resulting in greater pollution of the air.

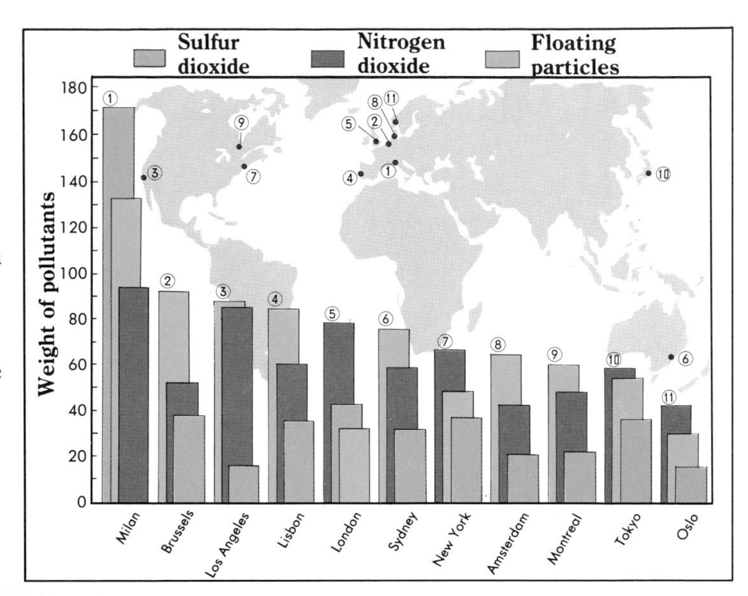


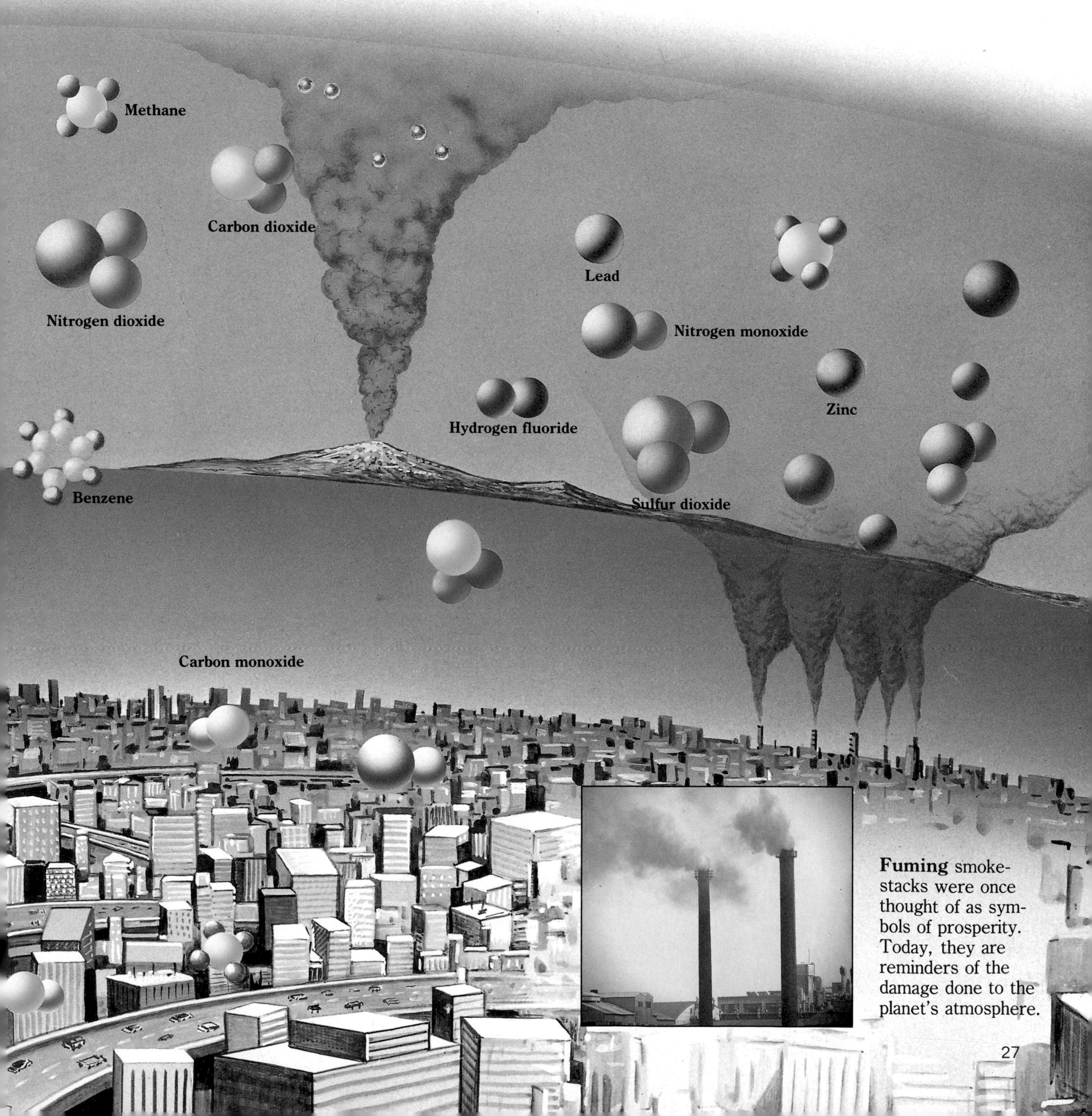


**Fuming** smokestacks were once thought of as symbols of prosperity. Today, they are reminders of the damage done to the planet's atmosphere.

# What Causes Acid Rain?

Even without air pollution, the carbon monoxide occurring naturally in the air makes rain slightly acidic. But in recent years, air pollution has increased the acidity of rain and snow to the point where precipitation has turned into a deadly threat to numerous organisms ranging from fish to forests.

When cars and factories burn fossil fuels, they release into the atmosphere pollutants such as sulfur and nitrogen oxides, halogen compounds, and a variety of hydrocarbons. These pollutants react with the moisture in the air to form highly acidic substances such as sulfuric acid, nitric acid, and hydrochloric acid. Raindrops and snowflakes then deliver these acids to rivers, lakes, and forests, making the soil and water inhospitable to life.

**Acid rain has killed** large portions of European forests such as the one above. Forests in the northeastern U.S. also have suffered heavy damage.

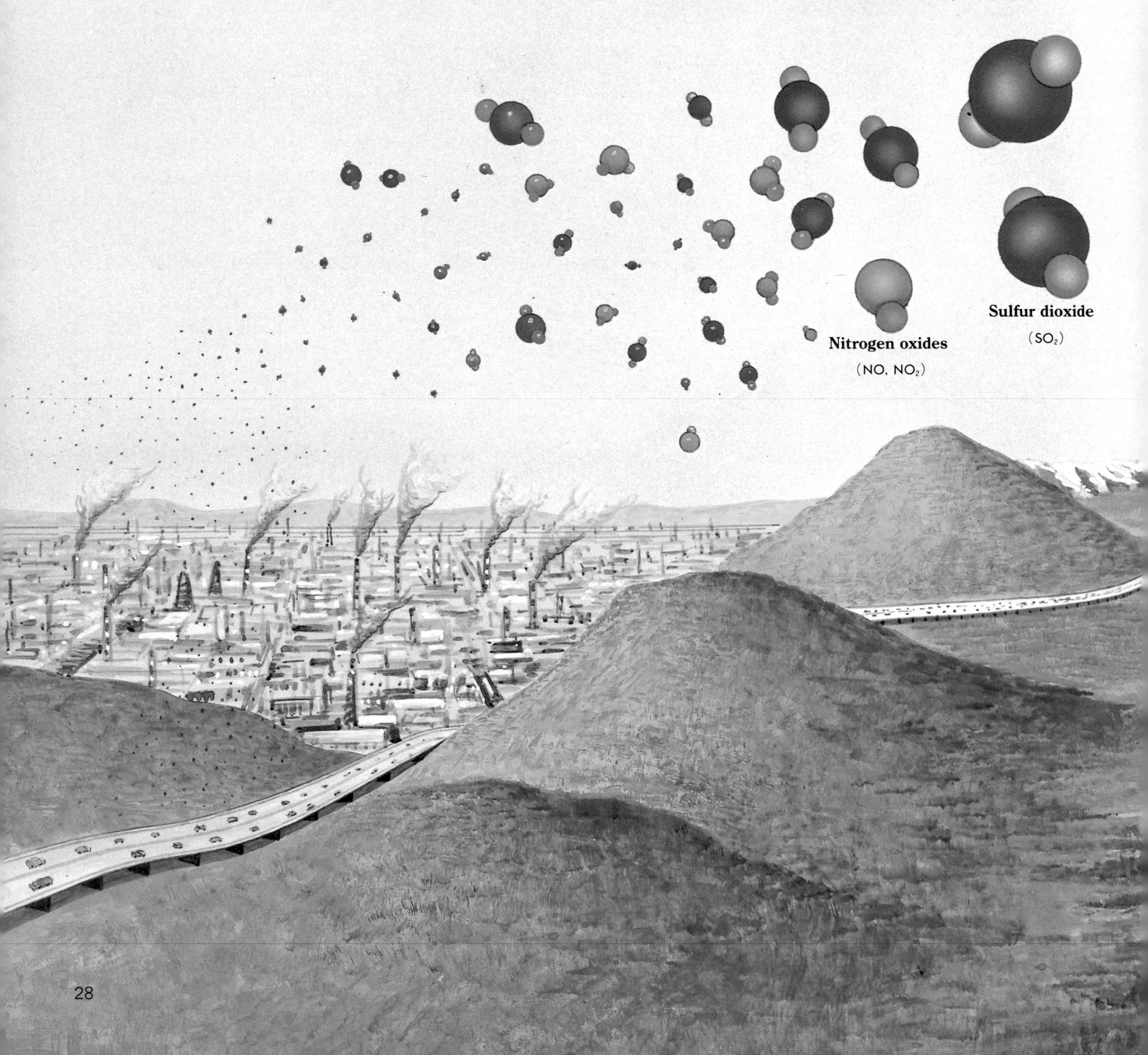

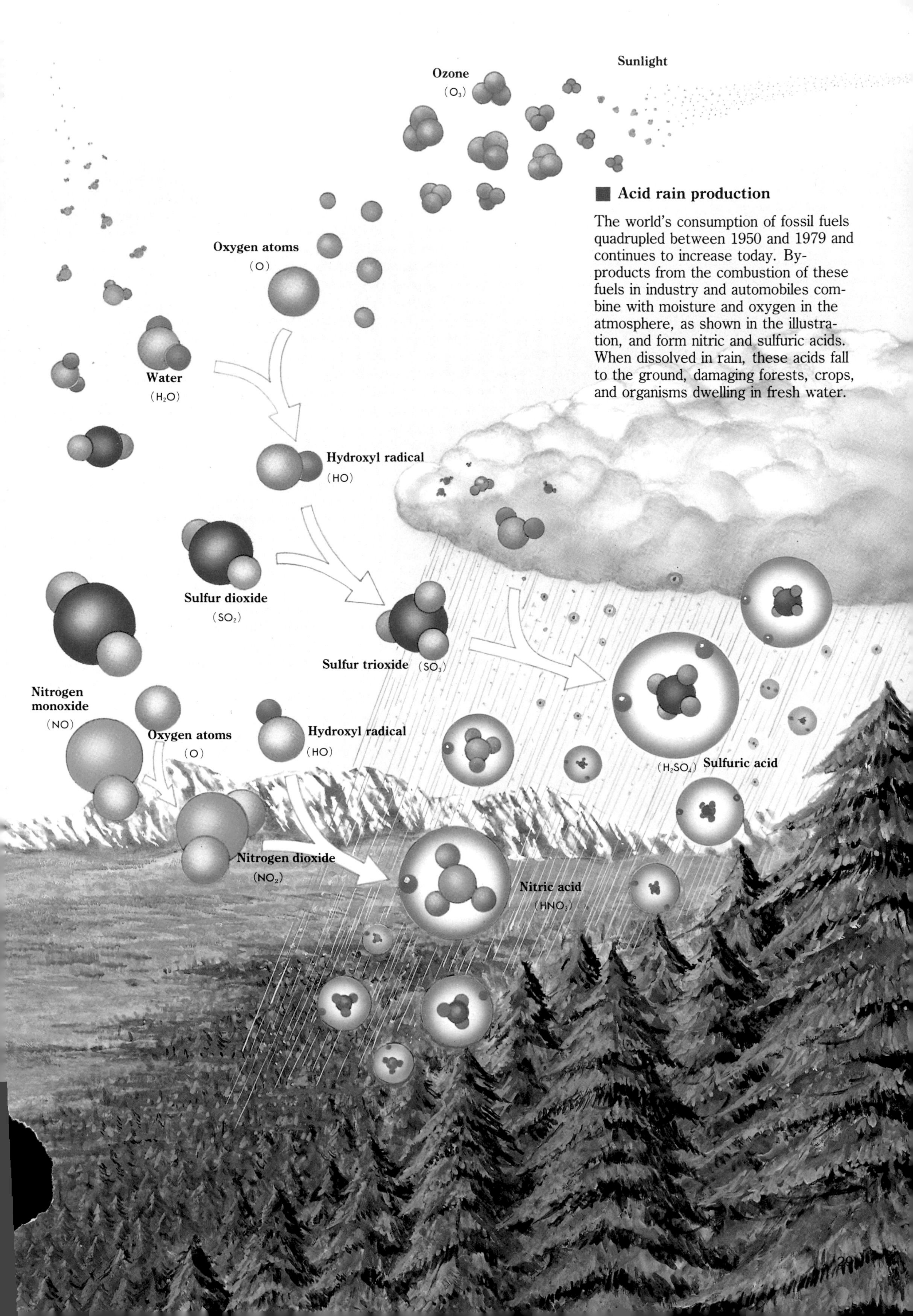


## Acid rain production

The world's consumption of fossil fuels quadrupled between 1950 and 1979 and continues to increase today. By-products from the combustion of these fuels in industry and automobiles combine with moisture and oxygen in the atmosphere, as shown in the illustration, and form nitric and sulfuric acids. When dissolved in rain, these acids fall to the ground, damaging forests, crops, and organisms dwelling in fresh water.

# What Are Ozone Holes?

In the mid-1980s, scientists were alarmed to discover that ozone in the atmosphere above the South Pole had been seriously depleted. Ozone, a triatomic form of oxygen, can play a harmful role in pollution when near the ground; however, in the stratosphere it forms an important barrier to harmful solar ultraviolet rays. Data from satellites showed that pockets of ozone-poor air, or ozone holes, were forming over the Antarctic continent in September and October, during the southern spring months. Scientists believed that chlorofluorocarbon (CFC) gases released as pollutants were destroying this ozone layer. The discovery of the ozone holes over the Antarctic led to a global ban on the production of CFCs that is to take effect by the year 2000.

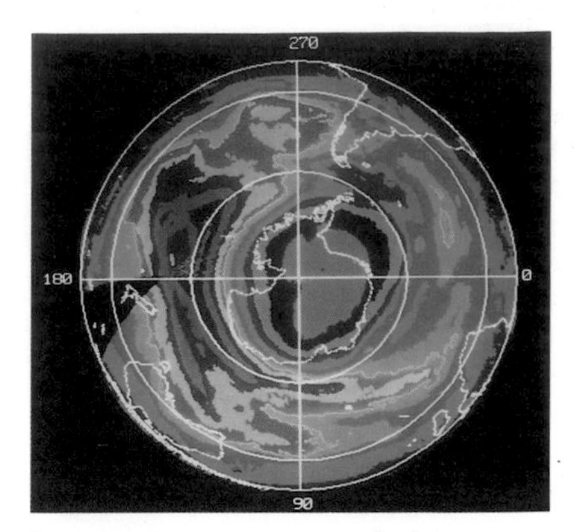


**A map** shows an ozone hole *(purple)* over Antarctica.

**1. Ozone is created** when ultraviolet rays break down oxygen molecules ($O_2$), which combine with free oxygen to form ozone ($O_3$). CFCs reaching the stratosphere are also broken down by UV rays, producing chlorine (Cl). The chlorine steals oxygen atoms from ozone, destroying the ozone and producing chlorine monoxide (ClO). The ClO reacts with $O_2$, eventually breaking apart to start the process over again.

**2. Above the Antarctic,** strong air currents develop during winter and spring, creating polar air cells that do not mix with outside air. The absence of sunlight in winter prevents production of new ozone inside the cells. Because of the extreme cold (as low as -112° F.), ice crystals form, letting the chlorine react with nitrogen compounds to form large amounts of hypochlorous acid.

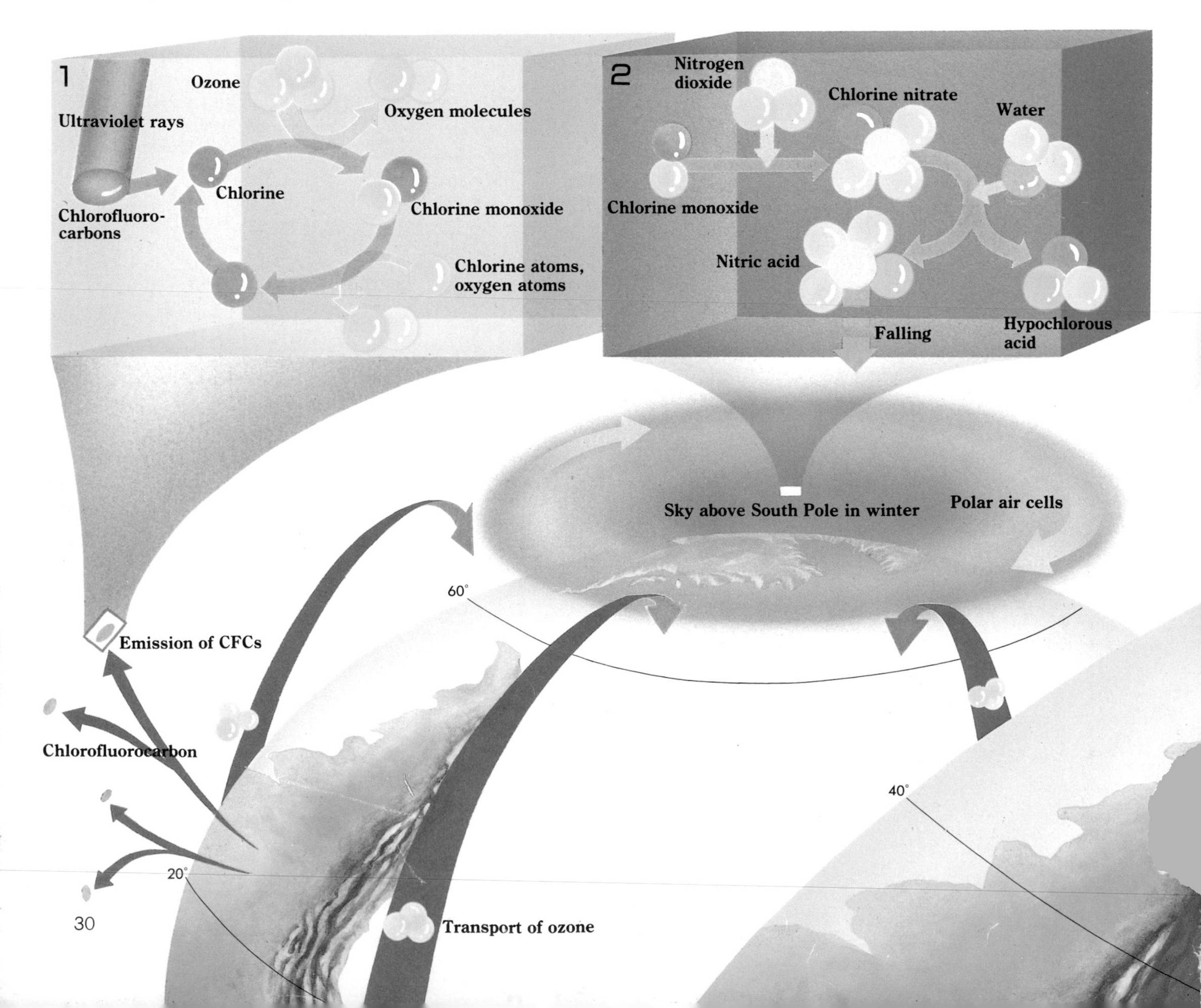

**3, 4. Springtime brings** destruction of the ozone above the Antarctic as sunshine returns and breaks down the hypochlorous acid formed during the winter. Chlorine is released and reacts, as shown below, to destroy the ozone. Atmospheric holes containing extremely low levels of ozone form over the Antarctic in this manner each spring. Similar holes could form over the populated Northern Hemisphere if CFC pollution continues.

## A graphic record

The graph at right records daily changes in ozone levels over the Antarctic in 1986. Ozone decreases during spring (September and October), but increases with summer's constant sunlight and renewed ozone production.

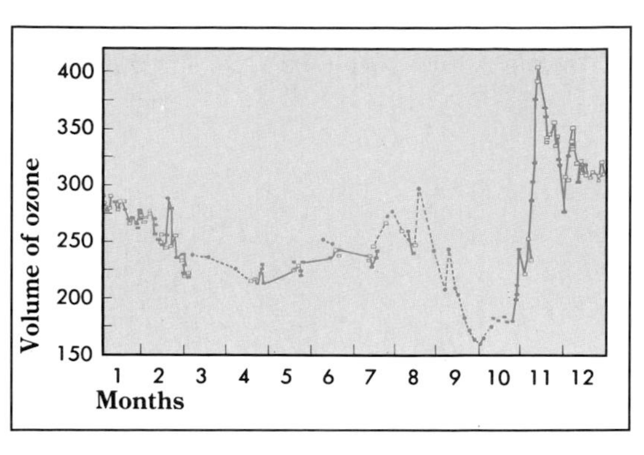


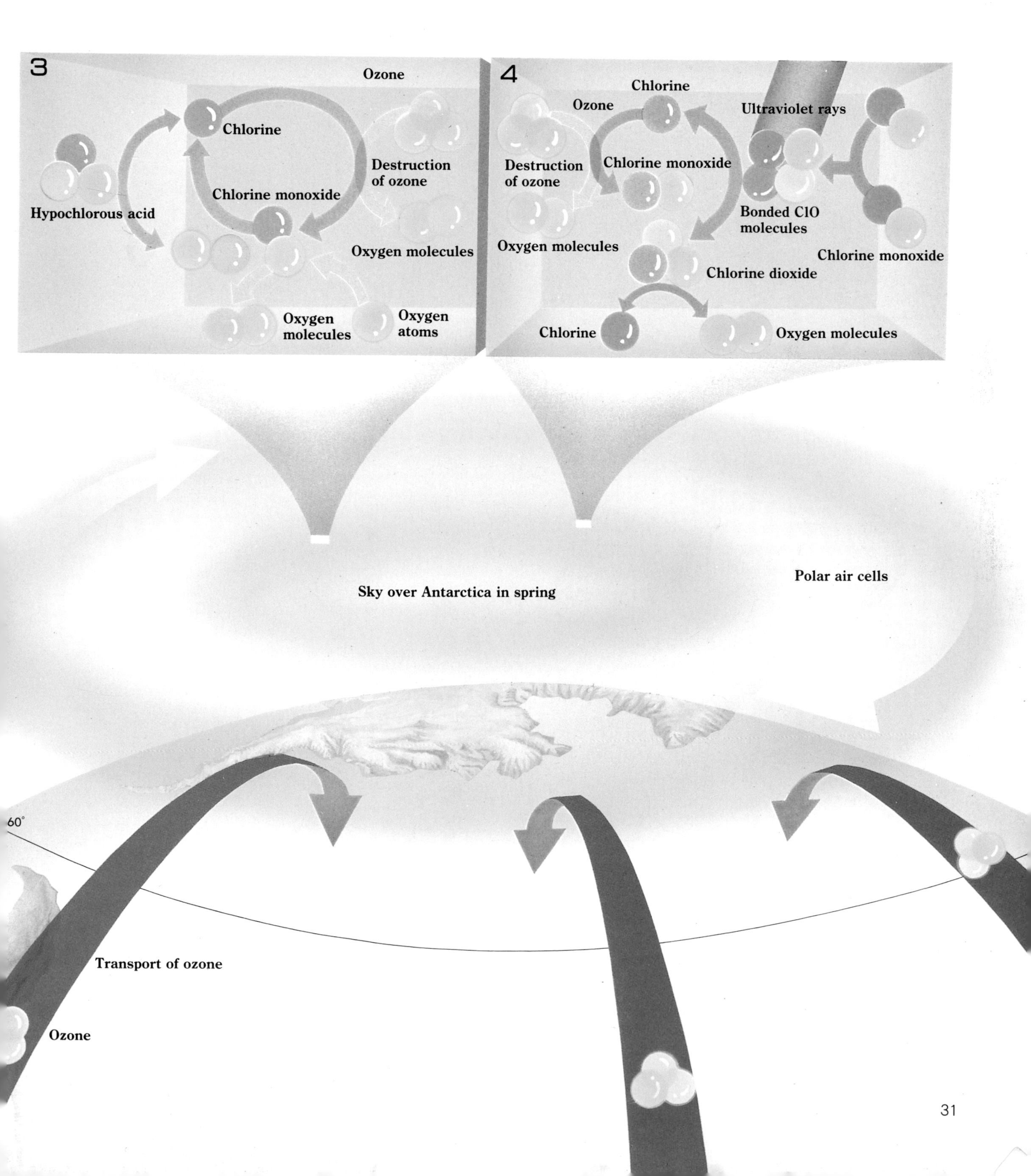

# 2

# The Air in Motion

There's not much to it, this stuff called air. It is what makes up the Earth's atmosphere, the blanket protecting life from the harshness of outer space—yet it has little substance. Only when the air is in motion is its presence felt.

A breeze drying laundry on the wash line, the mighty trade winds that cross the oceans, and the high-altitude jet streams that change the surface weather with every shift in course—all these are part of the Earth's atmosphere in motion. Powerful updrafts from the sun-warmed equator help to heat the rest of the planet and transport dust and sand around the globe. Winds moving over oceans carry moisture across the continents, creating rain and snow.

The air moves because the sun warms the Earth unevenly. The atmosphere receives little heat directly from the sun. Instead, the sun warms the Earth, which then radiates that heat into the air above it. As air becomes warm, it rises. This creates an area of low pressure beneath it that draws cooler air from a place where the Earth either did not absorb as much heat from the sun or did not radiate as much into the atmosphere. This can happen on many scales. On a local scale, air moving between a lake and its sandy beach creates a gentle breeze. On a global scale, air flowing from the hot equator to the frigid poles creates the general circulation of the atmosphere.

The most forceful winds are those found in a tornado *(right)*. These huge whirlwinds, common throughout the spring in the central United States, can be strong enough to pick up cars.

# Why Does the Atmosphere Circulate?

The Earth's atmosphere circulates in several giant loops. In the Ferrel cells, air moves along the surface toward one of the two poles and returns high in the atmosphere. In the other cells, the circulation is the reverse. This circulation exists because the sun's rays do not warm the Earth uniformly. At the North and South poles, the Earth loses more heat to outer space than it gains from the sun, while at the equator, the Earth absorbs more heat than it loses. As a result, the atmosphere is colder above the poles than at the equator.

If the Earth's atmosphere did not move, the air above the North and South poles would grow ever colder, while that above the equator would become hotter and hotter. What happens, though, is that warm air from the equator rises from the Earth's surface and moves toward the colder poles, forcing the colder air to move along the Earth's surface toward the equator. This process, called convection, transfers much of the excess heat from the equator toward the two poles and keeps the Earth's temperatures from becoming too unbalanced.

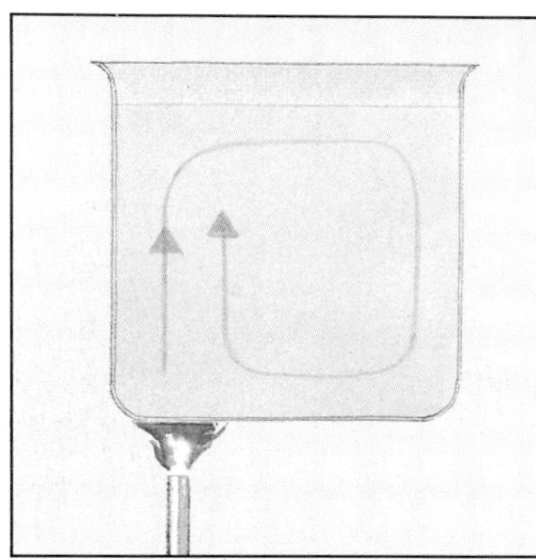

**Like** the heated atmosphere, heated water creates convection.

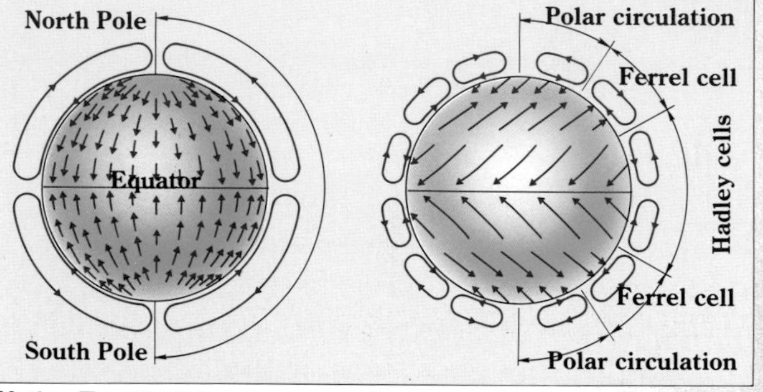


**If the Earth** did not rotate, the atmosphere would only circulate from the equator to the poles *(above, left)*. Instead, the Earth's spin produces six loops *(above, right)*.

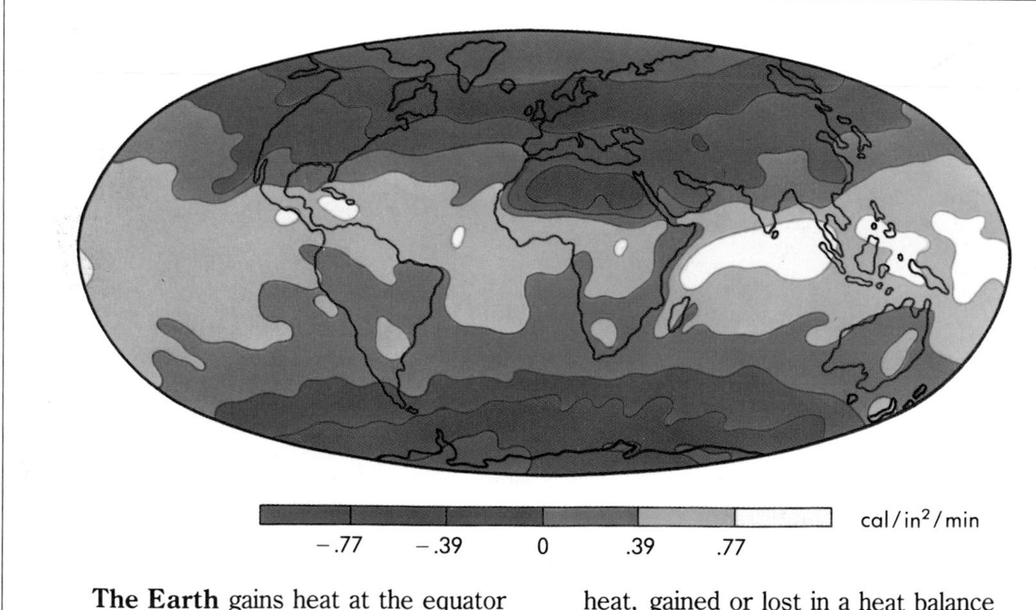


**The Earth** gains heat at the equator and loses heat at the poles, shown by the number of calories, or units of heat, gained or lost in a heat balance map *(above)*. Near 30° N and 30° S, the balance is close to zero.

## General circulation patterns

**Warm air at the equator** rises and moves toward the poles. When it reaches 30° latitude—north or south—it has cooled enough to sink back to Earth. Much of this air returns to the equator in the circulation pattern known as a Hadley cell, but the rest goes toward the poles. At about 60° latitude, this air collides with cold polar air flowing toward the equator. It then rises and doubles back toward the equator in a flow known as a Ferrel cell. The polar air, having absorbed heat from the Earth, also rises and goes back to its source, forming a polar circulation cell *(top and bottom).*

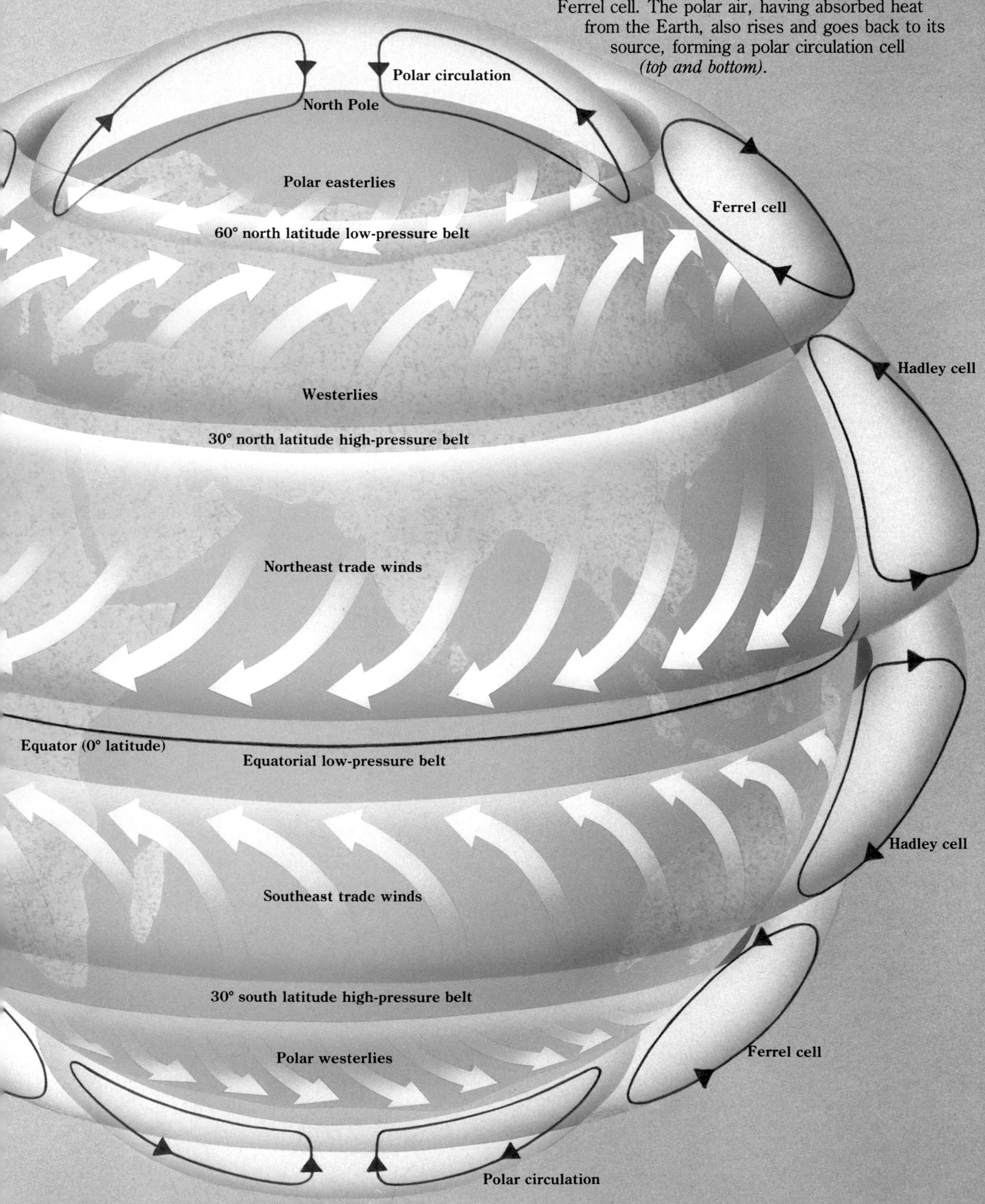

Polar circulation
North Pole
Polar easterlies
Ferrel cell
60° north latitude low-pressure belt
Hadley cell
Westerlies
30° north latitude high-pressure belt
Northeast trade winds
Equator (0° latitude)
Equatorial low-pressure belt
Hadley cell
Southeast trade winds
30° south latitude high-pressure belt
Polar westerlies
Ferrel cell
Polar circulation

# Why Are There Prevailing Winds?

Since ancient times, sailors have relied on prevailing winds to get them safely across oceans. These prevailing winds blow in a near-constant direction at certain latitudes. They arise from the general circulation of the atmosphere and from the Earth's rotation on its axis.

If the Earth did not spin on its axis, winds would move straight north or south. However, the Earth's spin generates a force of rotation called the Coriolis force, which deflects the winds. Because of this force, wind blowing northward or southward is deflected toward its right in the Northern Hemisphere and to its left in the Southern Hemisphere. Between 30° north latitude and 30° south latitude, wind blowing toward the equator twists to the west, producing the prevailing easterlies (blowing from east to west)—also known as the trade winds. By the same force, the wind that blows in the middle latitudes, moving toward the poles, is deflected eastward, becoming the prevailing westerlies (blowing from west to east). At the highest latitudes, the Coriolis force creates the prevailing polar easterlies.

**Rotation and wind direction**

Because the Earth does not stand still, the atmosphere does not move north and south, but rather in easterly and westerly directions. The Earth's spin deflects the path of wind to its right in the Northern Hemisphere and to its left in the Southern Hemisphere. In the Northern Hemisphere's Ferrel cell, for example *(middle row),* air moving north veers eastward.

**Prevailing polar easterlies**

**Earth standing still** | **Earth rotating** | **Wind direction**

**The prevailing westerlies**

**Earth standing still** | **Earth rotating** | **Wind direction**

**Trade winds**

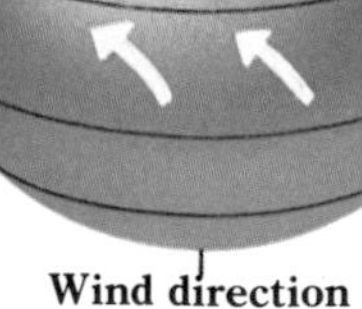

**Earth standing still** | **Earth rotating** | **Wind direction**

**Winds in stratosphere**

**Winds in thermosphere**

**Upper-atmosphere winds**

**Hadley cell**

9 mi

## General circulation and winds

Warm, buoyant air flowing toward the pole and cold, dense air returning to the equator create large cells in the general circulation of the Northern Hemisphere, seen in cross section below. Within these cells, air moving in the lower layers of the atmosphere produces the prevailing polar easterlies, prevailing westerlies, and trade winds. In the upper layers, air flow produces the jet streams. A strong ascending current occurs where the prevailing polar easterlies meet the prevailing westerlies of the Ferrel cell. This produces the low-pressure belt called the polar front. A high-pressure belt forms where the prevailing westerlies and trade winds meet.

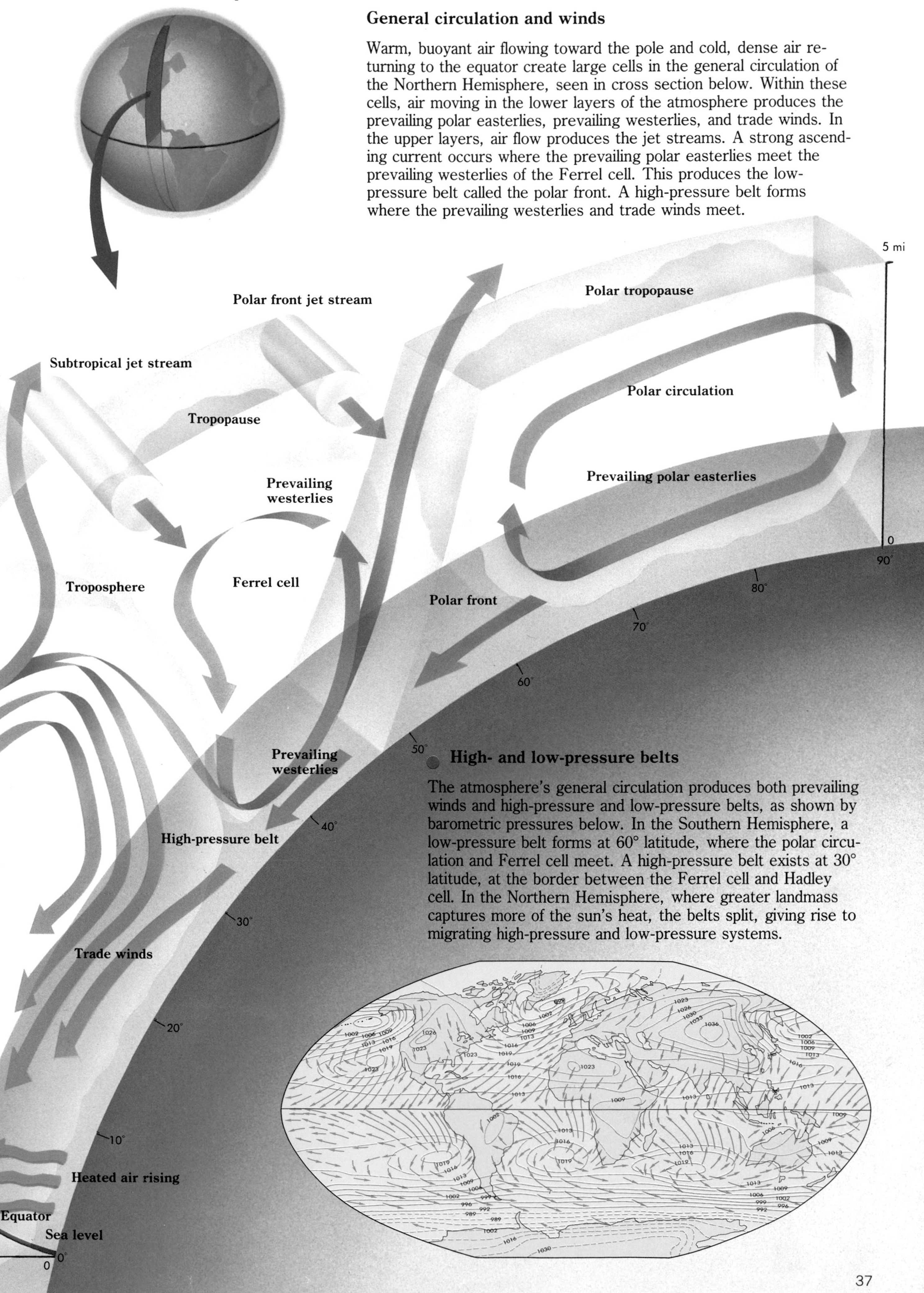


## High- and low-pressure belts

The atmosphere's general circulation produces both prevailing winds and high-pressure and low-pressure belts, as shown by barometric pressures below. In the Southern Hemisphere, a low-pressure belt forms at 60° latitude, where the polar circulation and Ferrel cell meet. A high-pressure belt exists at 30° latitude, at the border between the Ferrel cell and Hadley cell. In the Northern Hemisphere, where greater landmass captures more of the sun's heat, the belts split, giving rise to migrating high-pressure and low-pressure systems.

# What Causes the Jet Streams?

High above the Earth, at an altitude of 5 to 9 miles, powerful winds blow from west to east at speeds of up to 200 miles per hour. These are the jet streams. They form at the boundaries of the three major circulation cells in each hemisphere, where air masses of vastly different temperatures meet. These abrupt temperature shifts produce large pressure variations, which create strong winds. In winter, the temperature differences increase and the jet streams grow stronger.

During the northern summer, a reverse jet stream—blowing from east to west—forms over the Indian Ocean and Africa. The Asian landmass absorbs so much heat from the summer sun that the air above it becomes warmer than air over the equator. This hot, rising air creates a reverse jet stream, which produces India's monsoons.

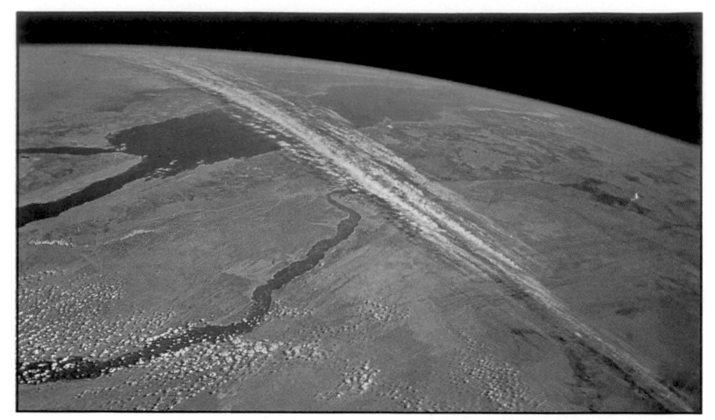

**Cirrus clouds over Egypt** mark the position of the subtropical jet stream in the Northern Hemisphere.

### Jet streams circle the Earth

Jet streams exist at the tropopause, where the troposphere meets the stratosphere. The polar jet stream forms at the boundary between the polar circulation and Ferrel cell. The subtropical jet stream forms where the Ferrel and Hadley cells meet.

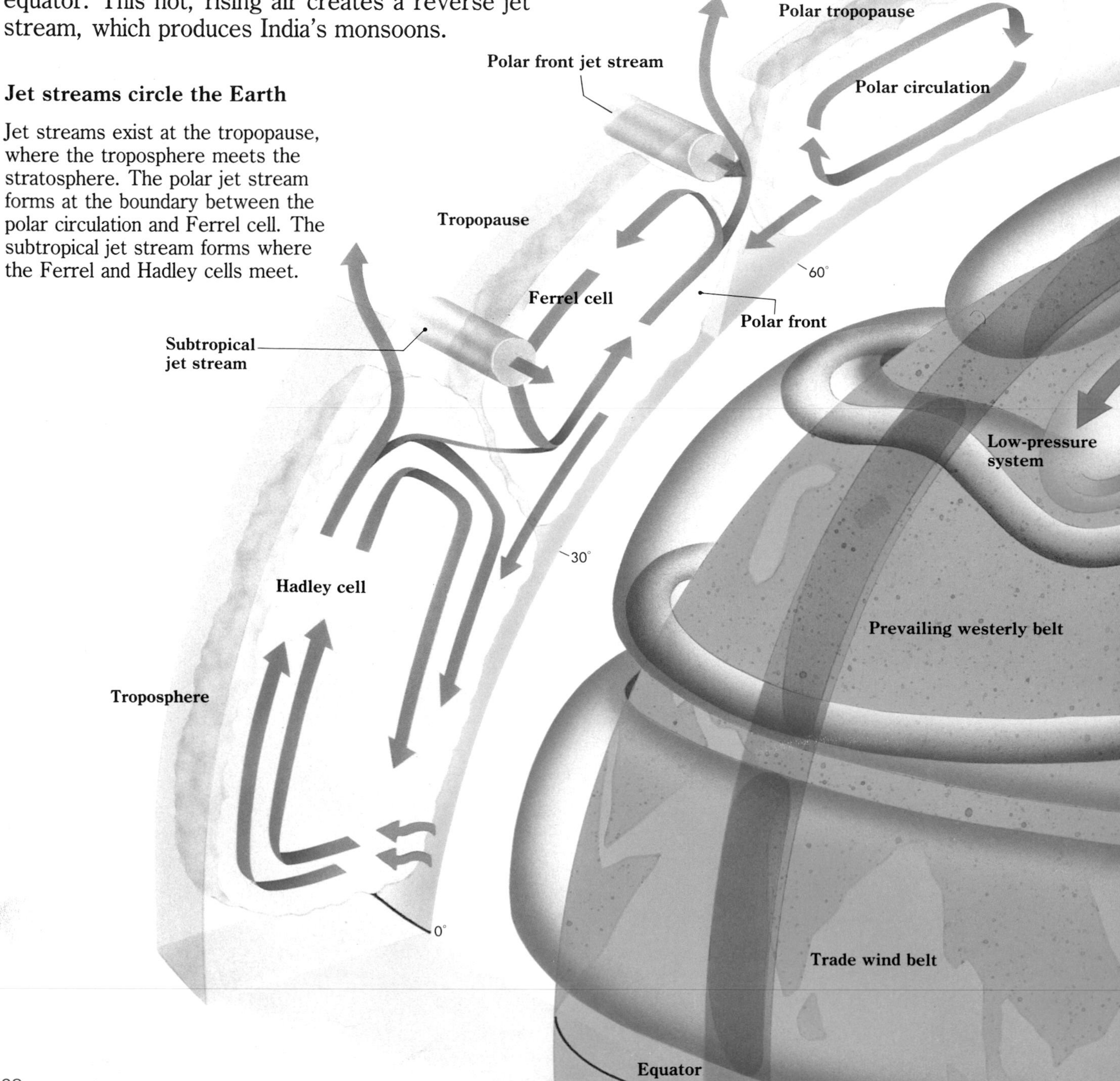

Winter

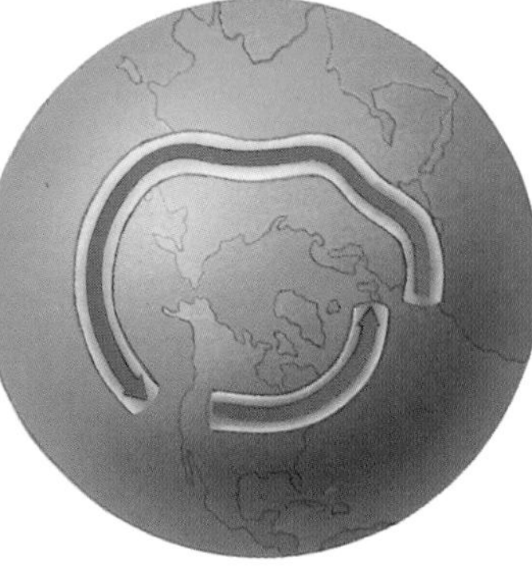

Summer

**The jet streams** meander north and south across the Earth with the changing seasons. In the Northern Hemisphere, the subtropical jet stream centers at about 30° latitude in January *(above, left)*.

By July, the jet stream has moved to about 40° latitude *(above, right)*. In winter, the subtropical jet stream blows harder and often merges with the polar jet stream, generating severe storms below.

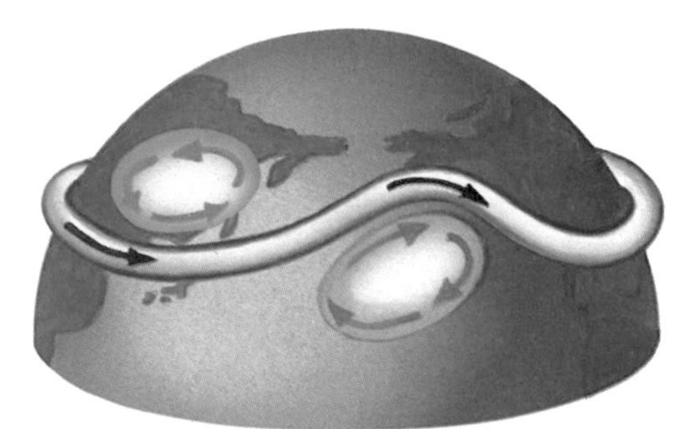

**High-pressure and low-pressure** systems deflect the polar jet stream *(above)*, making it bend like a moving snake. The subtropical jet stream zigzags less, because it encounters fewer pressure systems.

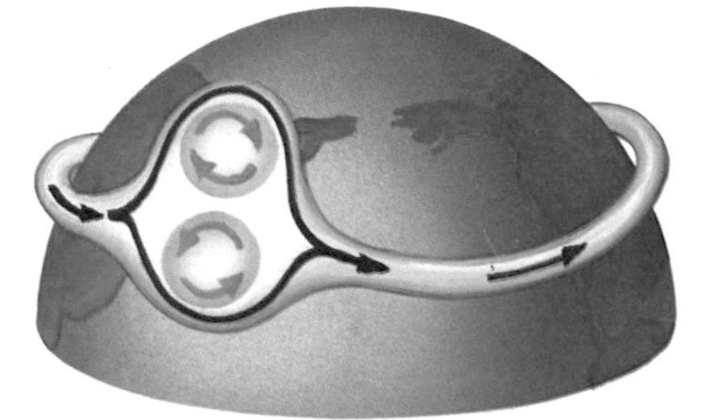

**The zigzag motion** of the polar jet stream can produce a high-altitude high-pressure system. A low-pressure system may then develop to the south, splitting the polar jet stream into two smaller jet streams.

North Pole

Prevailing polar easterly belt

Polar front jet stream

High-pressure system

Subtropical jet stream

High-pressure system

# Why Does Wind Blow?

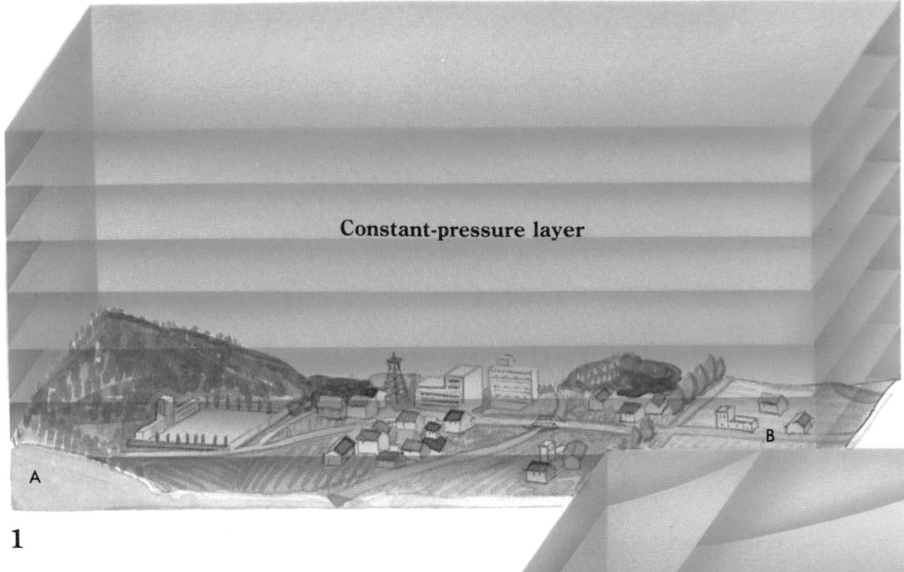


1

**Winds produced by convection**

Air pressure rises with temperature. Thus, when a warm air mass sits next to a cooler air mass, the two will have different pressures. This difference will set up convection currents *(Steps 1-4, below),* producing wind between the two zones.

**Equilibrium.** The temperature at points A and B *(left)* is the same, and so is the air pressure above them. No wind blows between the two points.

**Uneven warming.** The sun warms point B, raising the temperature of the air above it *(right).* The air expands and rises, and the air pressure increases.

2

Wind is air that is moving in relation to the Earth's surface, and it moves because of differences in air pressure in the atmosphere. Without these differences, no wind would blow. Differences in pressure develop over areas where the sun heats the Earth's surface unevenly. Wherever the Earth is warmer, air heats up and expands, and air pressure increases, compared to the pressure over cooler places.

Air can be imagined as lying over the Earth in layers between constant-pressure surfaces *(box, right),* with the densest layer at the bottom. Sometimes the air is still and the layers even and flat, as in Step 1. But when one area *(yellow, Step 2)* absorbs more heat, the air expands, air pressure rises, and the air pressure layers expand, too, and become curved.

Air then begins moving from the high-pressure area to the low-pressure area, producing a wind high above the ground, as in Step 3. The greater the temperature difference—and therefore the pressure difference—between two places, the stronger the wind that blows between them.

**Moving blocks of air**

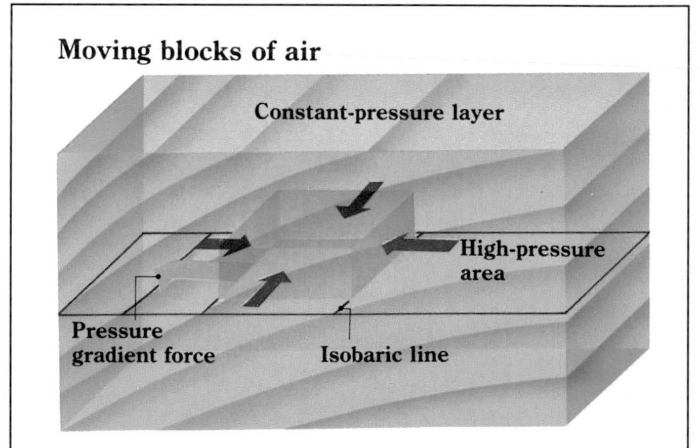


In drawing weather maps, meteorologists rely on a set of imaginary surfaces in the atmosphere, called constant-pressure surfaces *(curved planes, above).* All the points on such a surface have the same air pressure. Where an imaginary plane *(red outline),* parallel to the Earth, crosses a constant-pressure surface, scientists draw a line called an isobaric line, separating areas of differing air pressure. An air mass *(dark blue block)* between isobaric lines will be moved, by a barometric gradient force *(green arrow),* toward the lower-pressure area.

## Circular isobaric lines

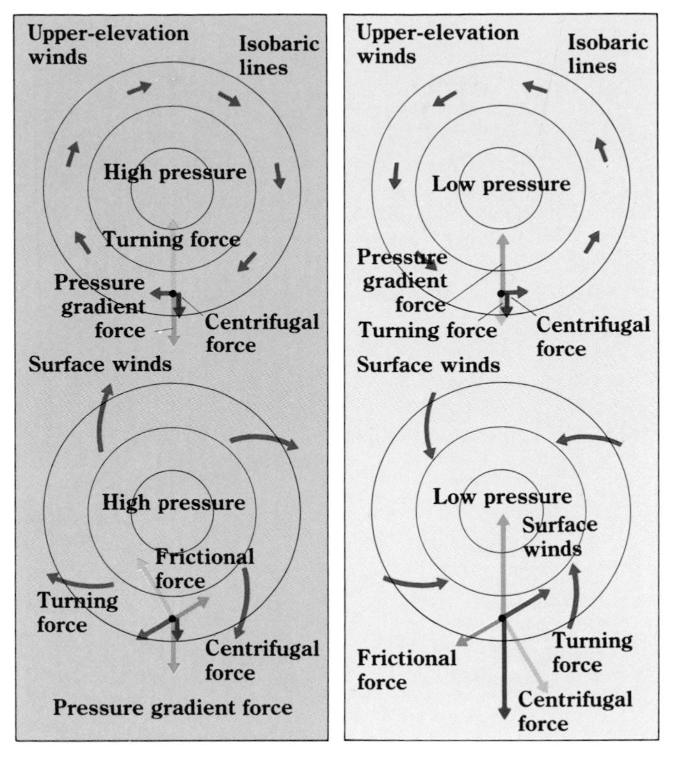


In high- and low-pressure systems, a centrifugal force also determines wind direction. In the upper atmosphere, the pressure gradient force, turning force, and centrifugal force balance when winds blow clockwise around high pressure *(far left, top)* and counterclockwise around low pressure *(near left, top)*. At ground level, friction turns the winds outward in a high *(far left, bottom)* and inward in a low *(near left, bottom)*.

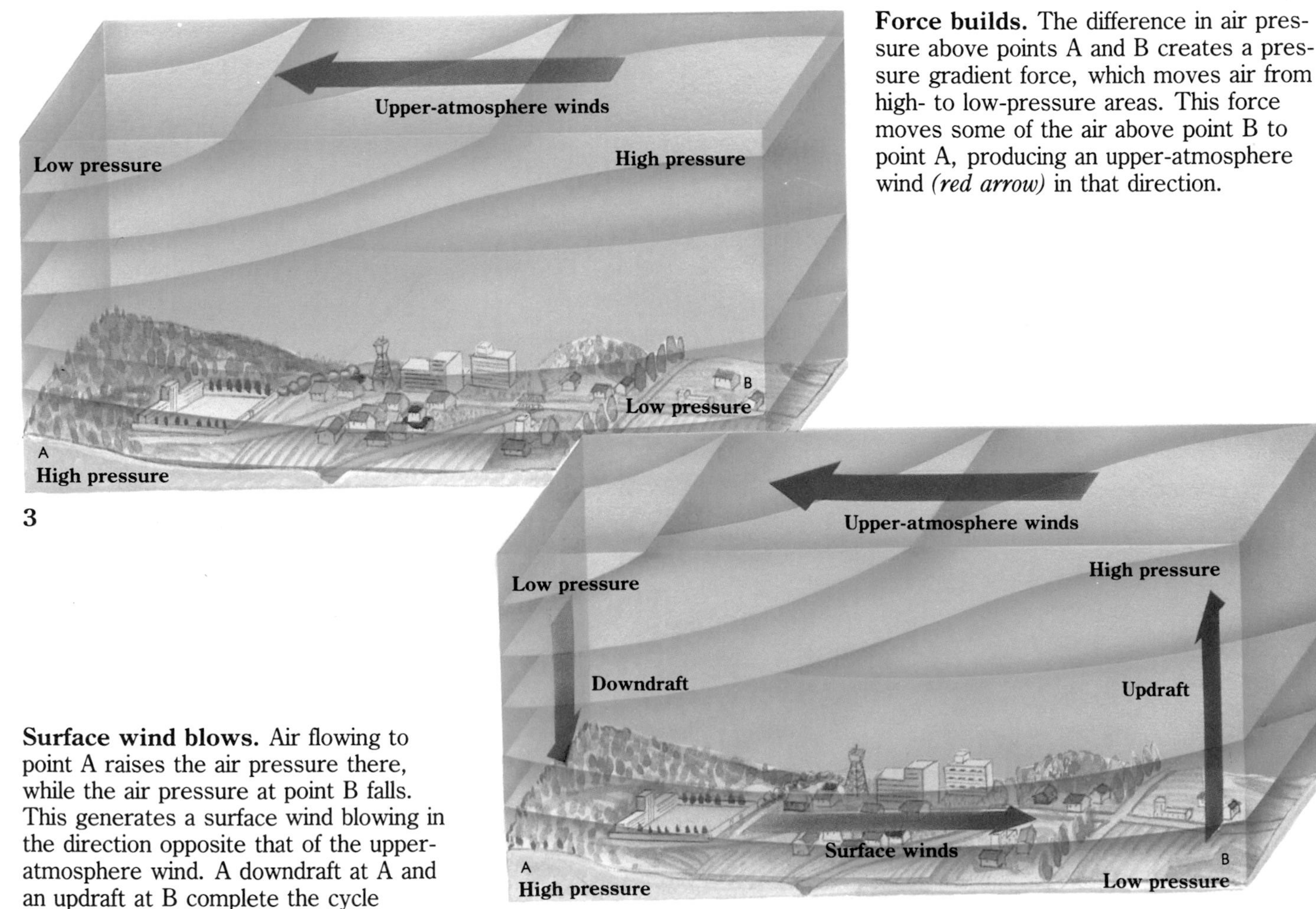


**Force builds.** The difference in air pressure above points A and B creates a pressure gradient force, which moves air from high- to low-pressure areas. This force moves some of the air above point B to point A, producing an upper-atmosphere wind *(red arrow)* in that direction.

**Surface wind blows.** Air flowing to point A raises the air pressure there, while the air pressure at point B falls. This generates a surface wind blowing in the direction opposite that of the upper-atmosphere wind. A downdraft at A and an updraft at B complete the cycle

# What Causes Regular Breezes?

Winds are fairly unpredictable, but some breezes blow as regularly as the rising and setting of the sun. Unlike most winds, these breezes depend more on local changes in temperature than on the presence of large-scale high- and low-pressure systems. Near an ocean or large lake, for example, sea and land breezes blow because the sun warms land faster than it does water. By the same token, land loses heat faster than water when the sun stops shining.

In a valley, the air rises and sinks over the course of one day as the surrounding mountainsides gain and lose heat. The updrafts and downdrafts that are created by this temperature change produce valley and mountain breezes that blow virtually every day the sun is out.

**Land and sea breezes**

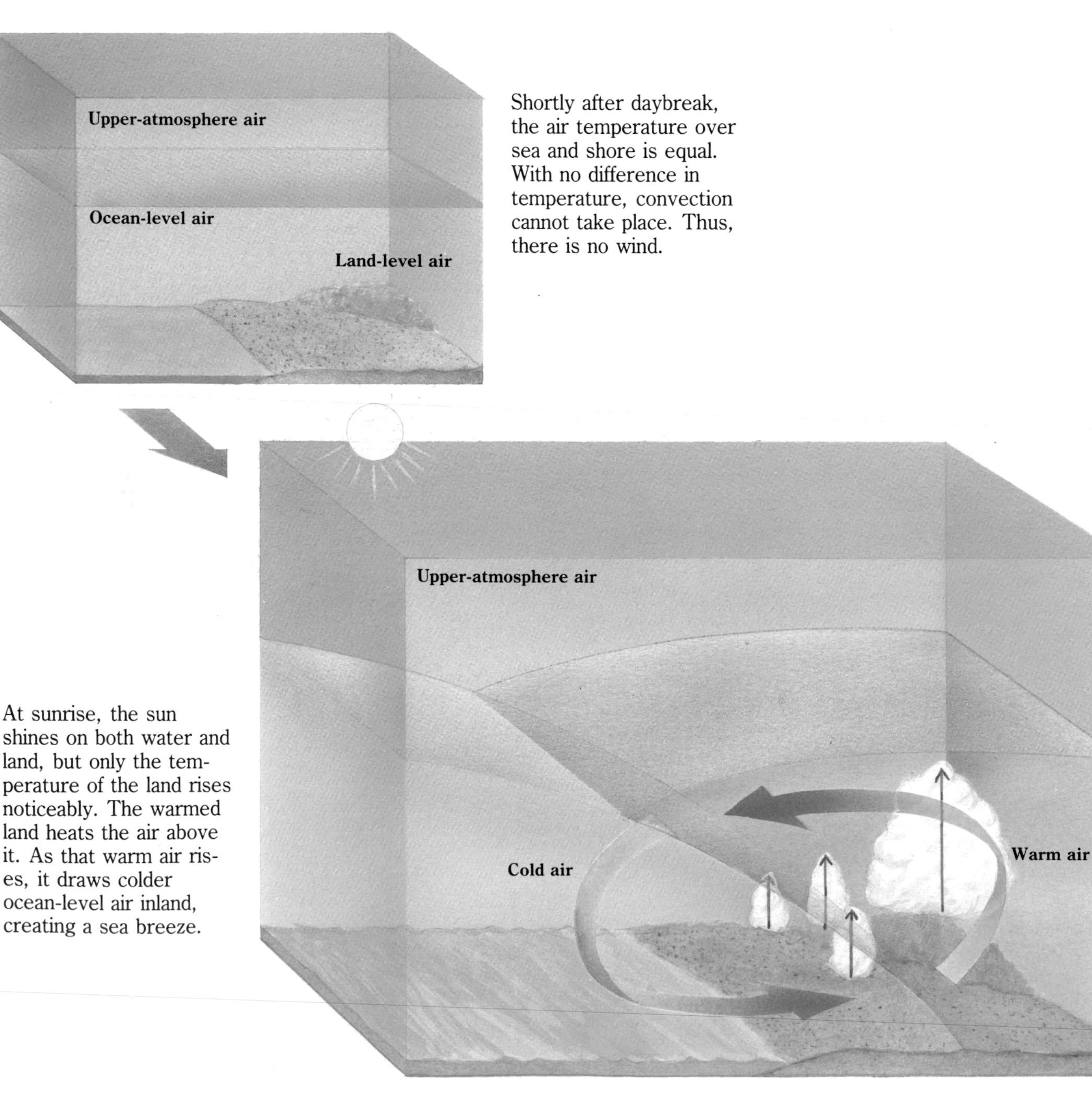


Shortly after daybreak, the air temperature over sea and shore is equal. With no difference in temperature, convection cannot take place. Thus, there is no wind.

At sunrise, the sun shines on both water and land, but only the temperature of the land rises noticeably. The warmed land heats the air above it. As that warm air rises, it draws colder ocean-level air inland, creating a sea breeze.

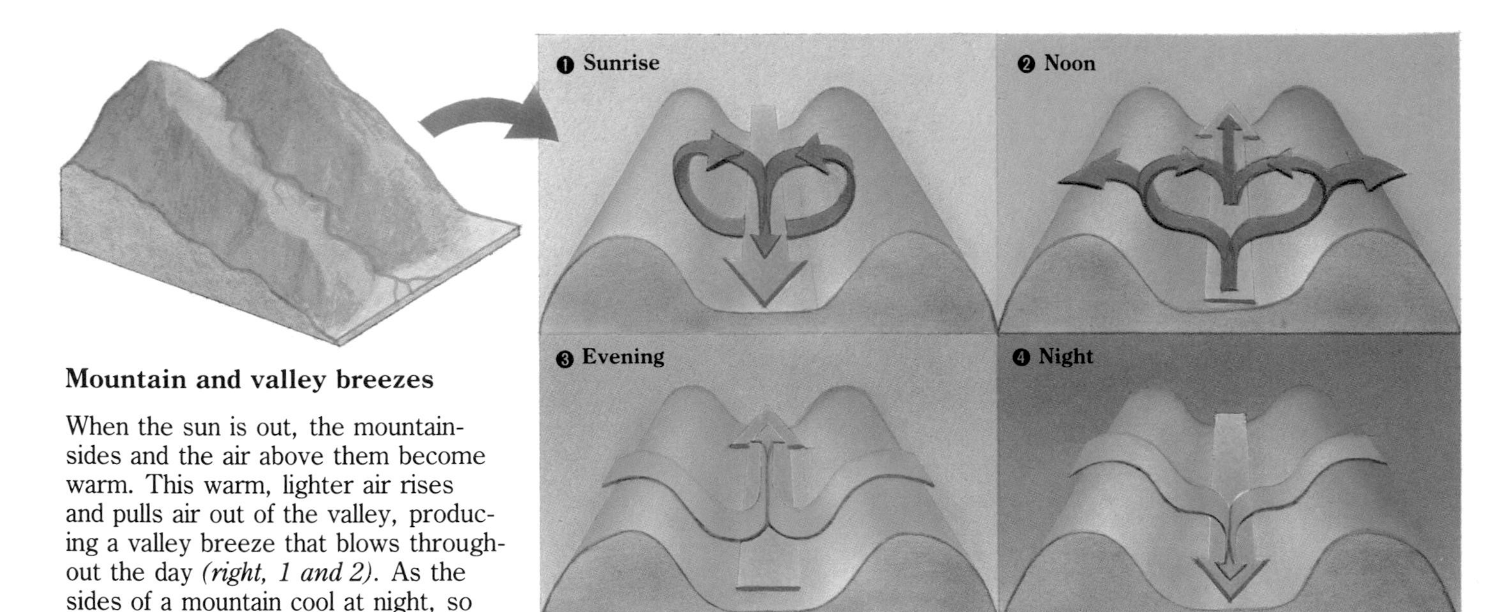


### Mountain and valley breezes

When the sun is out, the mountainsides and the air above them become warm. This warm, lighter air rises and pulls air out of the valley, producing a valley breeze that blows throughout the day *(right, 1 and 2)*. As the sides of a mountain cool at night, so does the air above them. This cooler, denser air then sinks into the valley, creating a mountain breeze that blows all night *(right, 3 and 4)*.

Cool mountain air is flowing into the valley as the sun rises *(1)*, but by noon the valley has warmed enough to reverse the breeze. At night, the breeze changes direction once more as cool air again flows into the valley *(4)*.

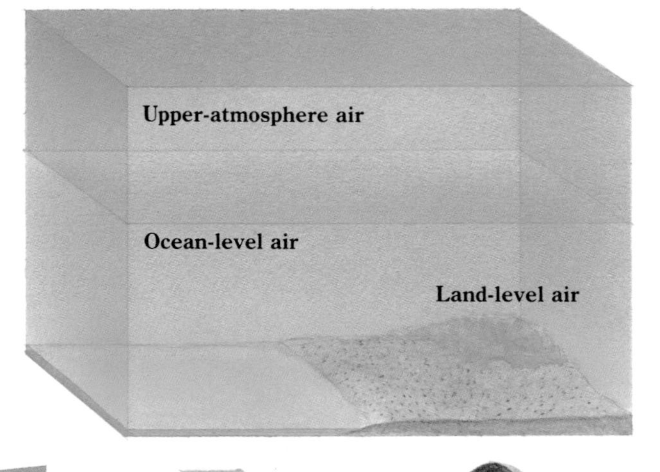


**Temperature changes.** This graph shows how land and sea surface temperatures change on a sunny day. Though sea surface temperature remains nearly constant, land temperature can vary by 50° F. over the course of a day. The times where the temperature lines intersect are when the morning and evening calms occur.

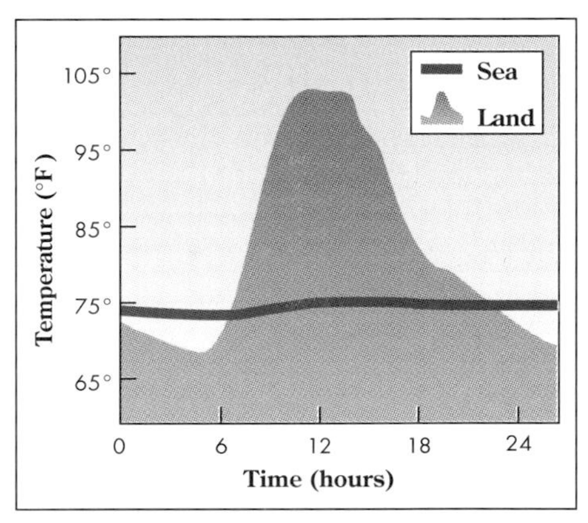


**The evening calm.** Once the sun sets, the shore begins cooling and so does the air above it. Before long, the air temperatures over the ocean and shore are the same. The sea breeze stops blowing. This is the evening calm.

**Nighttime.** As night progresses, the shore becomes cooler than the surface of the water. The air over the ocean is then warmer compared to that over land, so it rises. Cooler air from shore then moves out toward the sea, creating a land breeze.

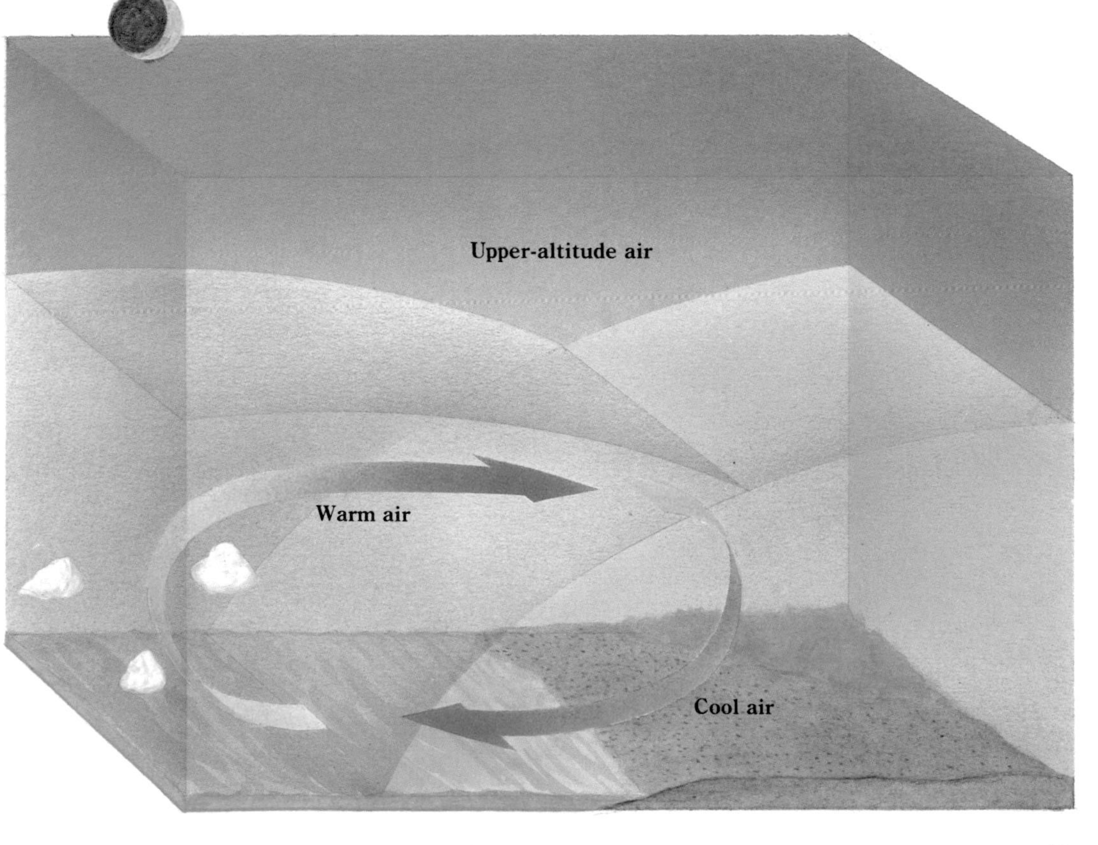

# What Brings Warm Winter Winds?

One of the strangest winds is the foehn, a hot, dry wintertime blast that blows down the slopes of many of the world's iciest mountains. Foehns, which include the Rocky Mountain chinook, begin forming when warm air rises from lowlands to mountain peaks. As it rises, the air loses much of its moisture as rain or snow. After crossing the mountain peak, the air moves rapidly downward, becoming compressed and warmer. When it reaches the foot of the mountain, the air can be 30° F. warmer than when it crossed the summit.

A foehn can be a blessing and a curse. In Switzerland, its warmth ripens fruit in the fall and lets Rhine Valley farmers grow corn and grapes. In the western United States, chinooks often save cattle from starvation by uncovering grasses buried under snow. But a foehn can also melt snow so fast as to cause severe flooding.

**Rocky Mountain chinook**

When a large low-pressure system sits on the western slopes of the Rocky Mountains, it draws warm, moist air from the Pacific Ocean over the mountains *(box on globe at right)*. The air cools as it climbs the mountain, losing its water-carrying capacity. Clouds form, and rain or snow falls. After crossing the summit, the air spills down the mountainside, becoming compressed as it goes. This heats the air, producing the warm chinook.

Chinooks blow all along the Rocky Mountains, from Alberta, Canada, to northwestern New Mexico. Typically, a chinook covers an area 180 to 300 miles wide, and can raise temperatures by 20° to 40° F. in just a few minutes. In Havre, Montana, the temperature once rose from 11° F. to 42° F. in three minutes. Chinooks have been known to devour a 10-inch snow cover overnight.

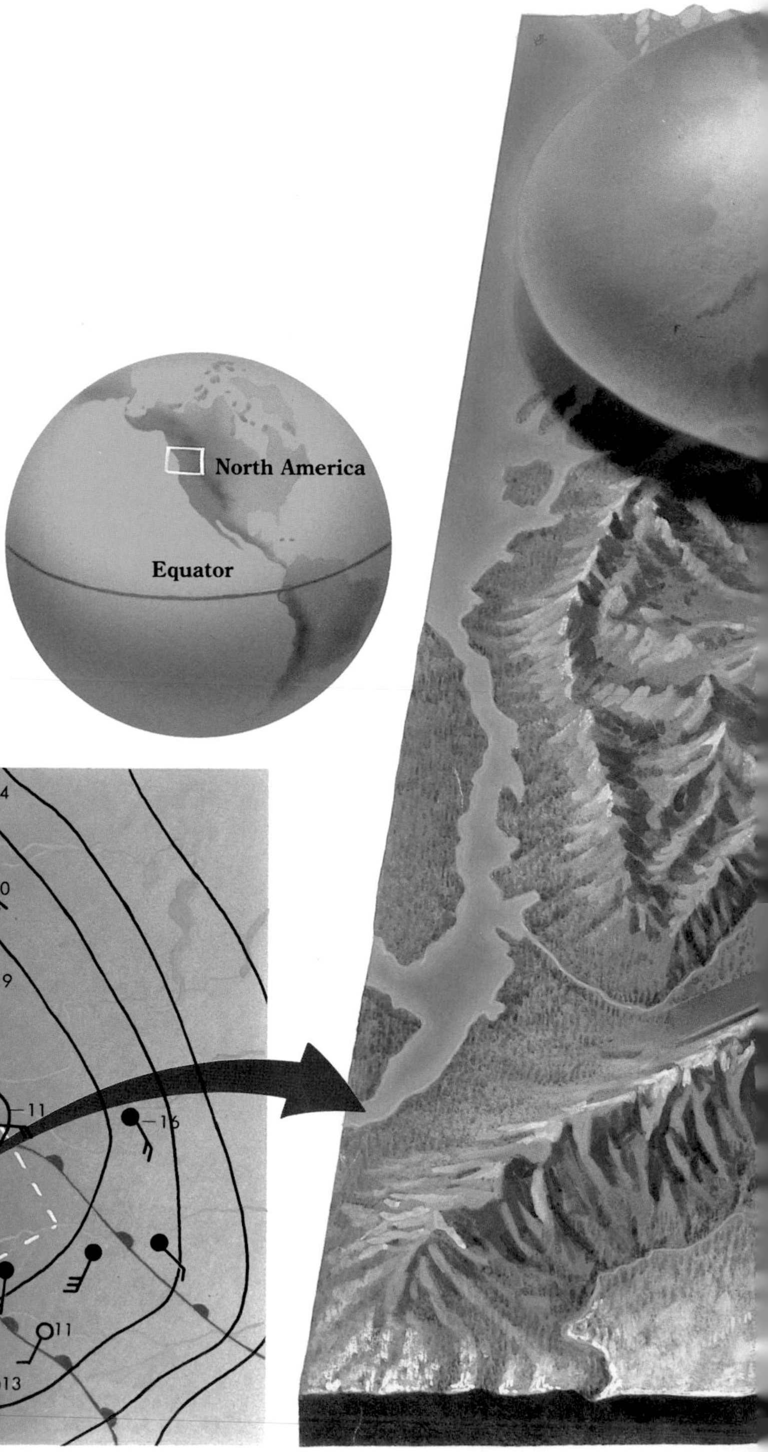

**Chinooks bring** dramatic temperature changes. At 11:00 p.m. on a February night, the temperature during this Rocky Mountain chinook reached 13° C. (55° F.).

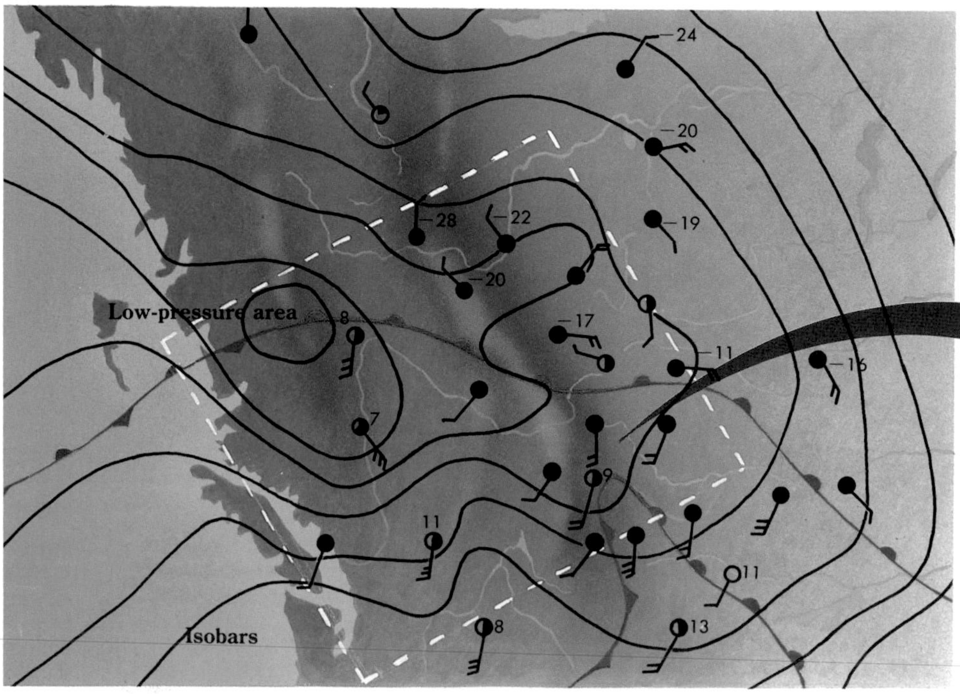

## Anatomy of a foehn

As warm, moist air climbs the slopes of a mountain, it cools and its relative humidity rises *(left scale)*. When the air becomes saturated with moisture, clouds form, slowing the rate at which the air cools. After crossing the summit, the air begins falling down the mountain. This compresses the air, raising the temperature by 5.5° F. every 1,000 feet it descends *(right scale)*. Warm, dry air flows onto the plain below.

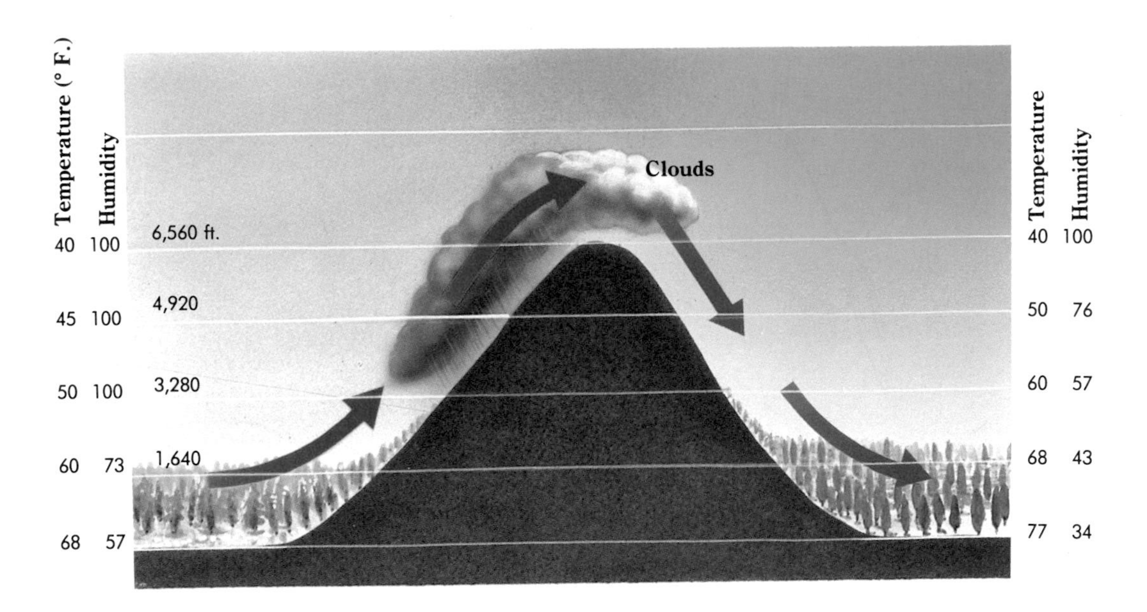


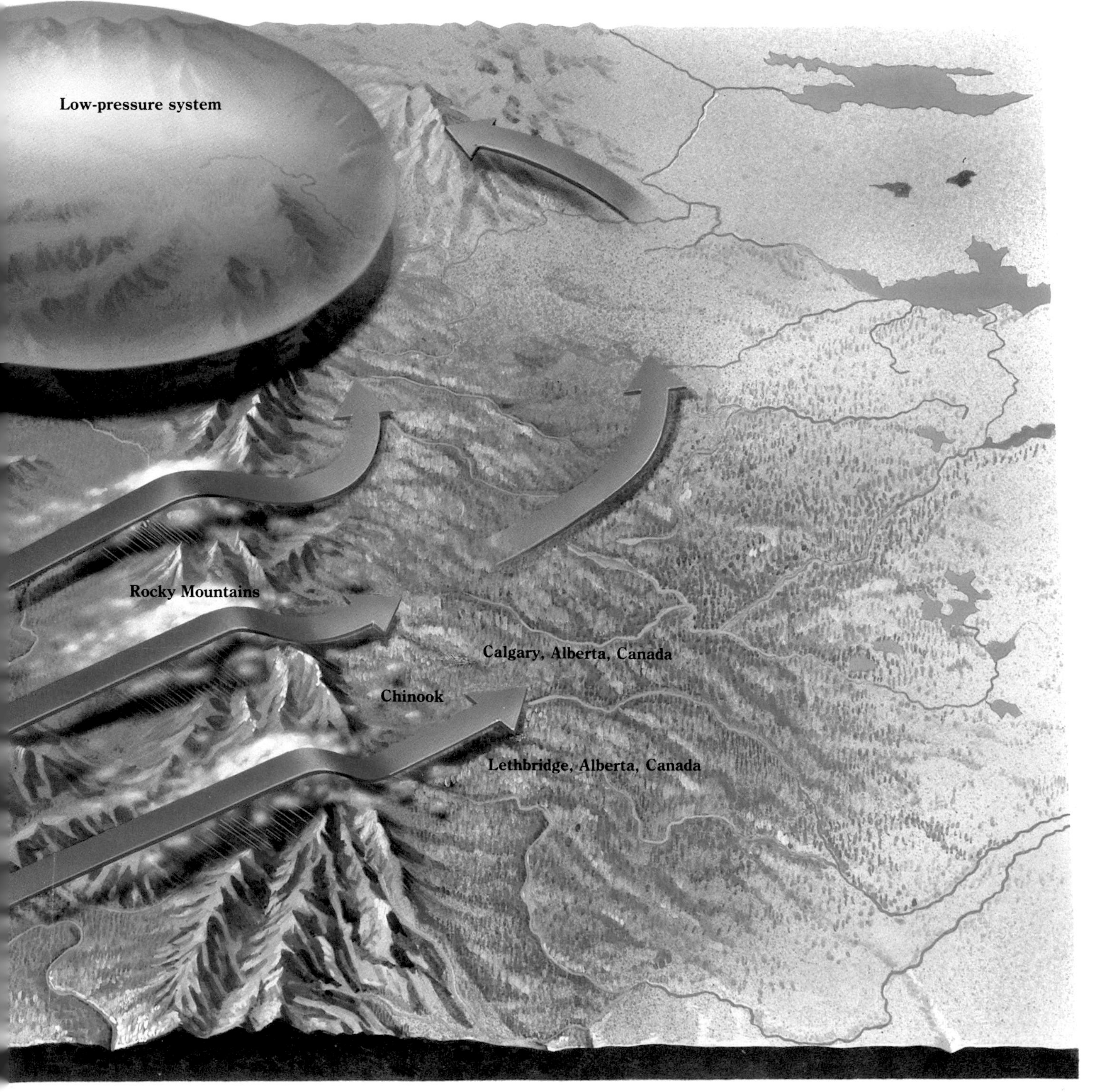

# What Is the Sirocco?

The Mediterranean basin forms a highway for low-pressure systems created in the Atlantic Ocean. These low-pressure systems draw air from over the Sahara, creating the hot, sand-laden wind known in southern Europe as the sirocco. When the sirocco blows and the sun shines, temperatures can exceed 100° F. Since the sirocco also picks up moisture from the Mediterranean on its way to Europe, the weather it brings is usually muggy as well as hot. Rain that falls during the sirocco is often discolored by the large amounts of desert sand and dust in the air.

**Clouds form** as a moist sirocco climbs steep Mediterranean peaks.

**Low-pressure highway**

Strong winds swirl in toward the low-pressure systems that drive into the Mediterranean basin from the Atlantic Ocean. The most powerful of these winds, the sirocco, comes from the south, carrying sand and heat from the Sahara. The sirocco blows most commonly in the spring, when air over the Sahara is hot and the low-pressure systems entering the basin are strongest.

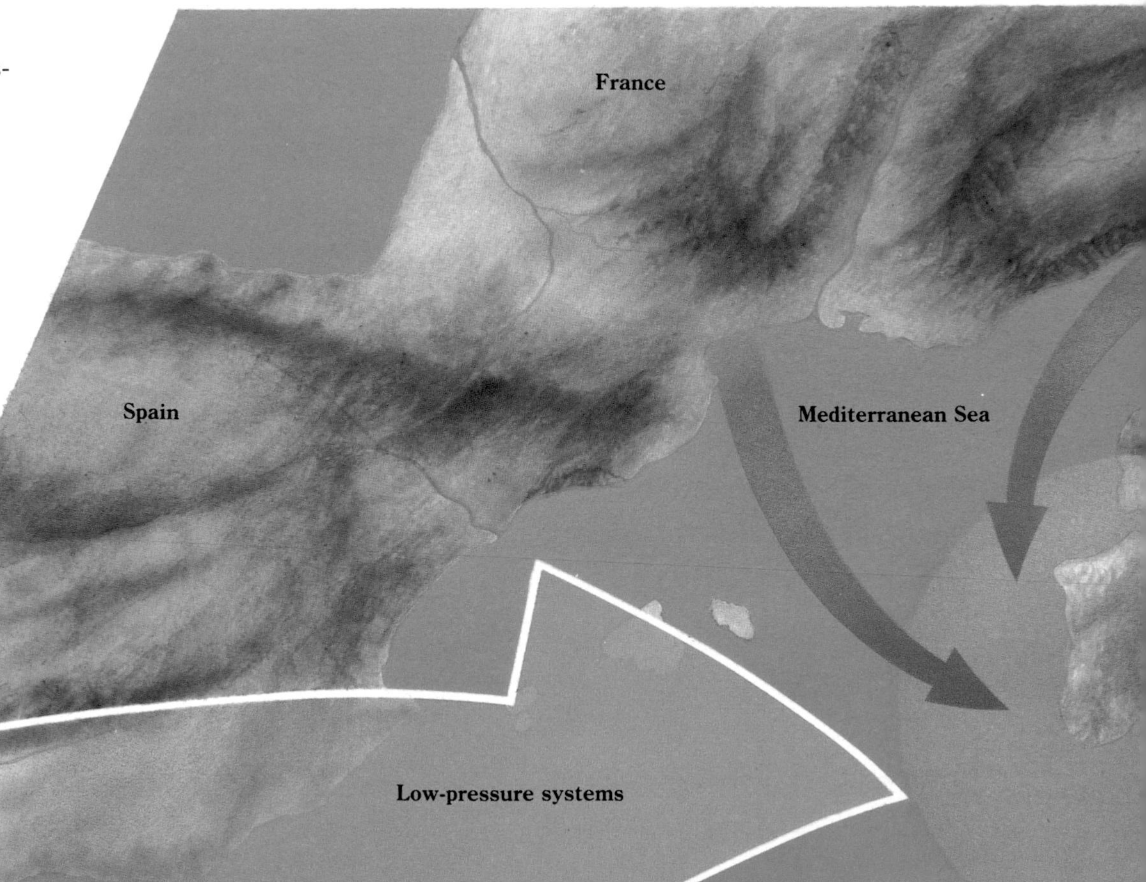

## The foehn in Sicily

The sirocco picks up moisture as it crosses the Mediterranean *(map and inset below)*. When this wet air ascends Sicily's mountains, clouds form *(inset)* and rain falls. The dry air then goes down the other side of the mountain, compressing and heating up. The foehn keeps Sicily's southern side humid and lush and its northern slopes dry and barren.

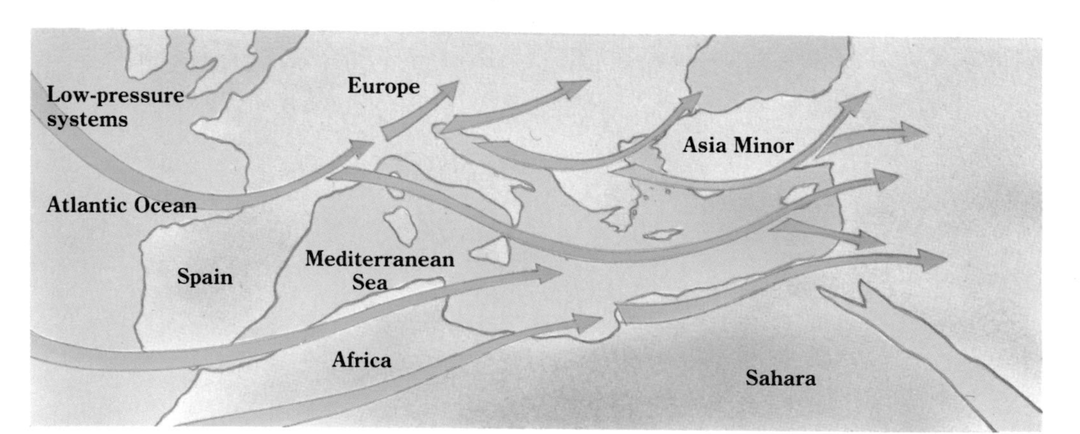


**Southern Europe, the Mediterranean,** and northern Africa form a basin that provides a natural path for low-pressure systems from the Atlantic Ocean.

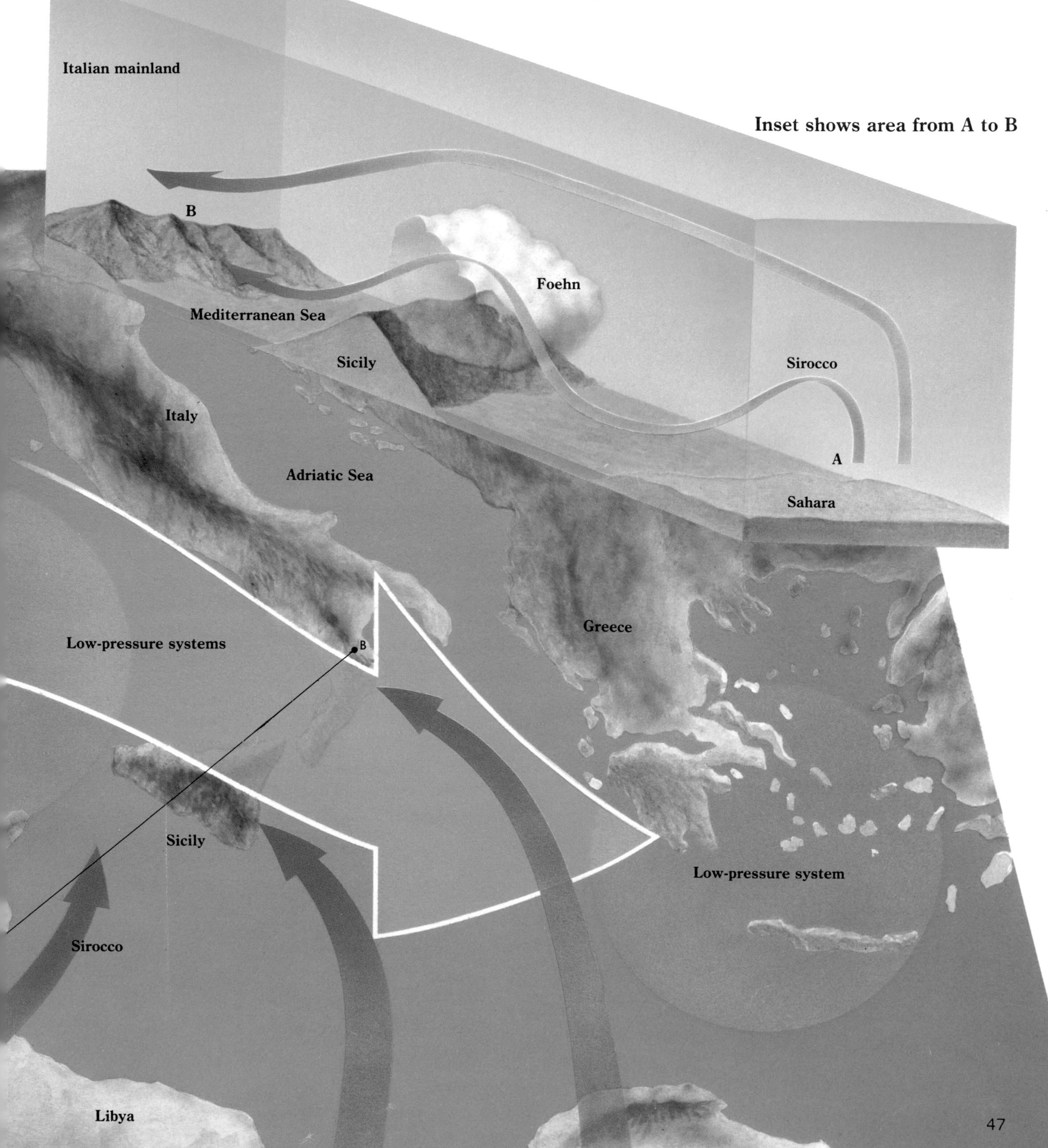

# Why Does the Adriatic Bora Blow?

**1 High pressure** develops over Yugoslavia's plateaus. High pressure brings clear skies, so heat escapes into the atmosphere, chilling the air mass. The Dinaric Alps, 4,000 to 8,000 feet high, block this cold air from reaching the Adriatic.

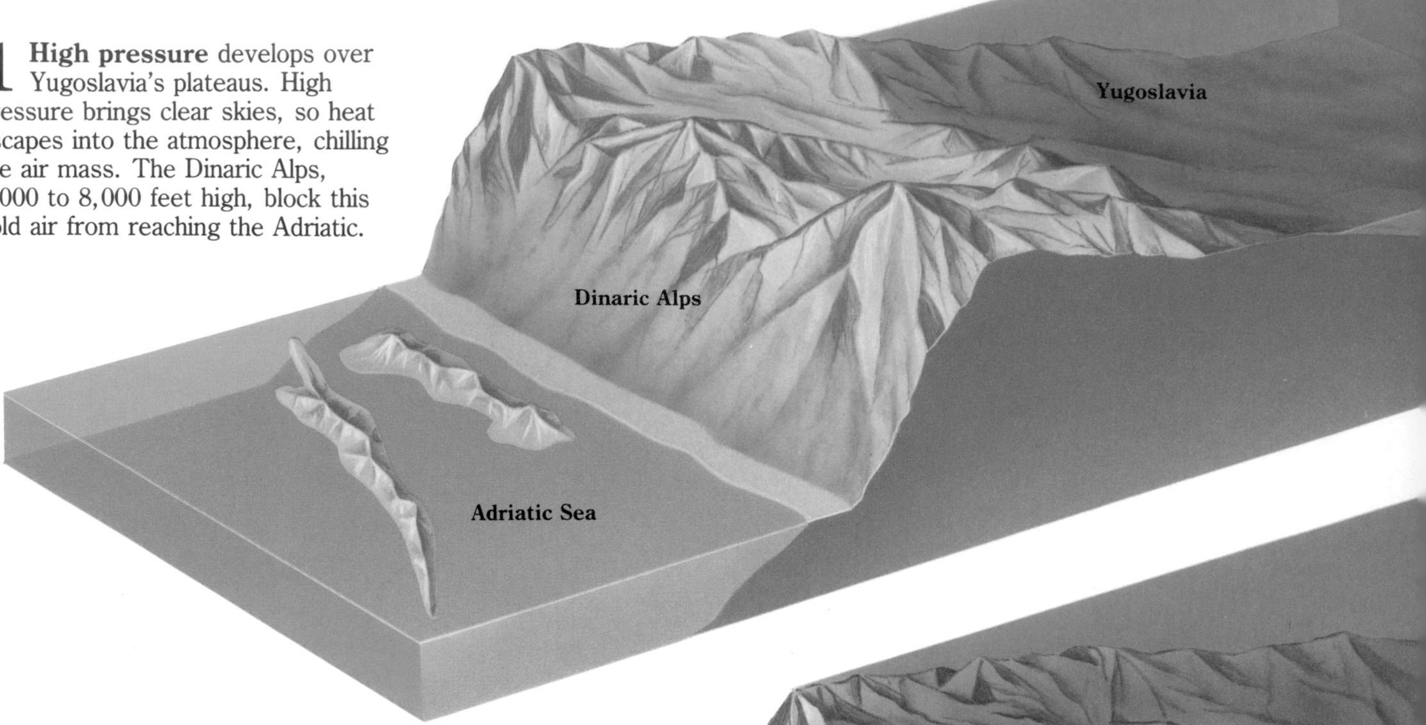

**2 A low-pressure system** moves into the Adriatic Sea, drawing the trapped cold air up over the Dinaric Alps. The bora winds blow strongest through a 50-mile depression in the mountain range.

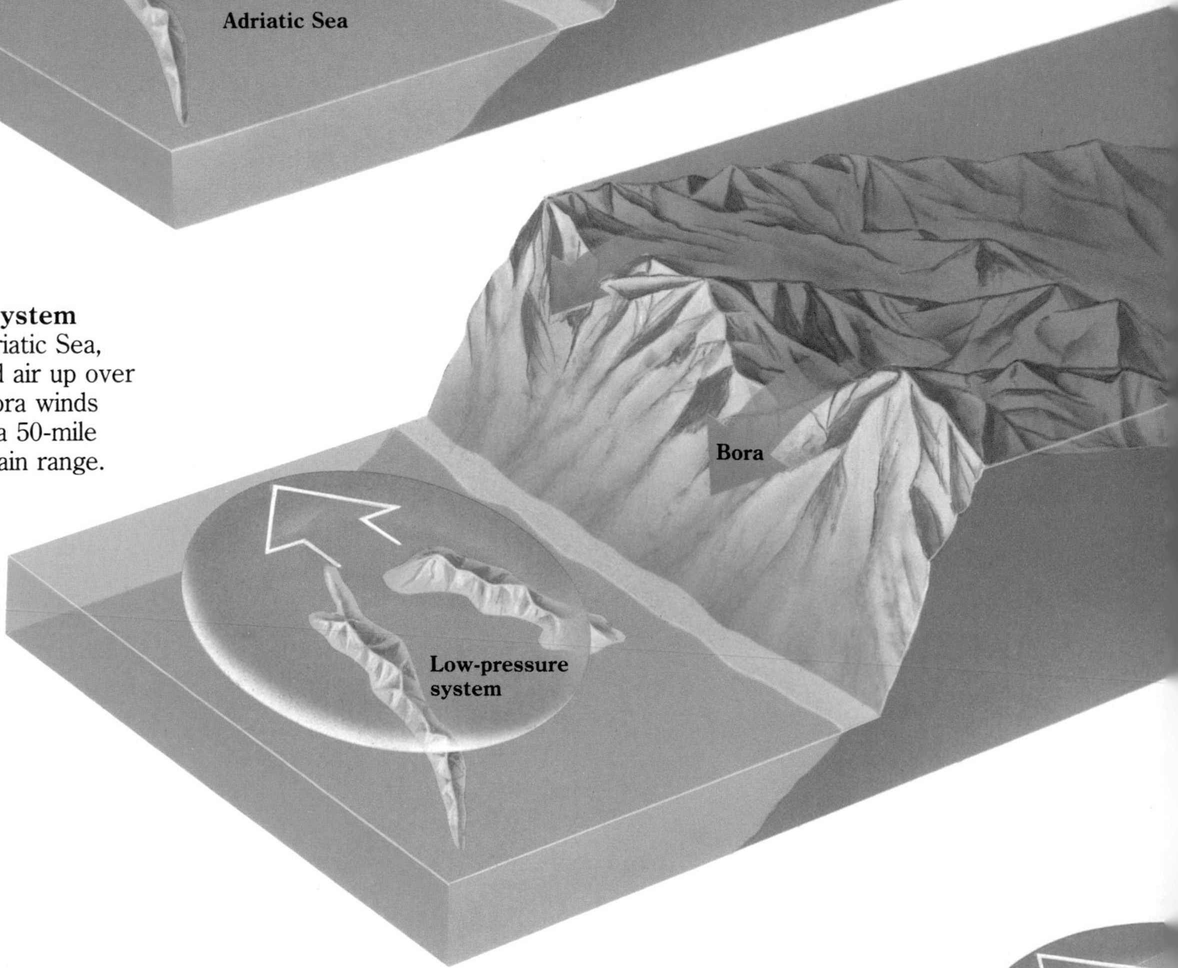

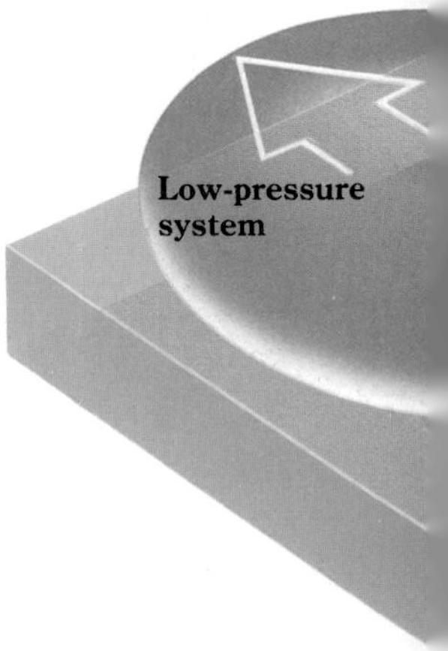

The eastern shore of the Adriatic Sea is known for its warm winters, which it owes to mild southerly breezes from the Mediterranean Sea. On occasion, though, frigid air flows over the Dinaric Alps, bringing the cold, dry wind known as the bora.

The bora begins when a cold air mass collects in the Yugoslavian plateau, trapped by the peaks of the Dinaric Alps. Low pressure, swinging north into the Adriatic on its journey across the Mediterranean, can draw this trapped air over the mountains and down onto the towns along the coast. When the bora blows, the usually warm, moist air quickly becomes cold and dry. Sometimes, the bora even brings snow and ice, making travel unexpectedly treacherous.

## Where the bora blows hardest

Yugoslavia's Dinaric Alps *(left)* reach as high as 8,000 feet, trapping cold continental air behind their peaks. However, in the 50 miles between Crikvenica and Senj, the mountains are a mere 4,000 feet high. This dip in the mountain ridge provides the easiest passage for the trapped air to reach the sea, and it is through here that the bora blows particularly hard.

**The Yugoslavian city** of Dubrovnik often faces the bora's frigid blast.

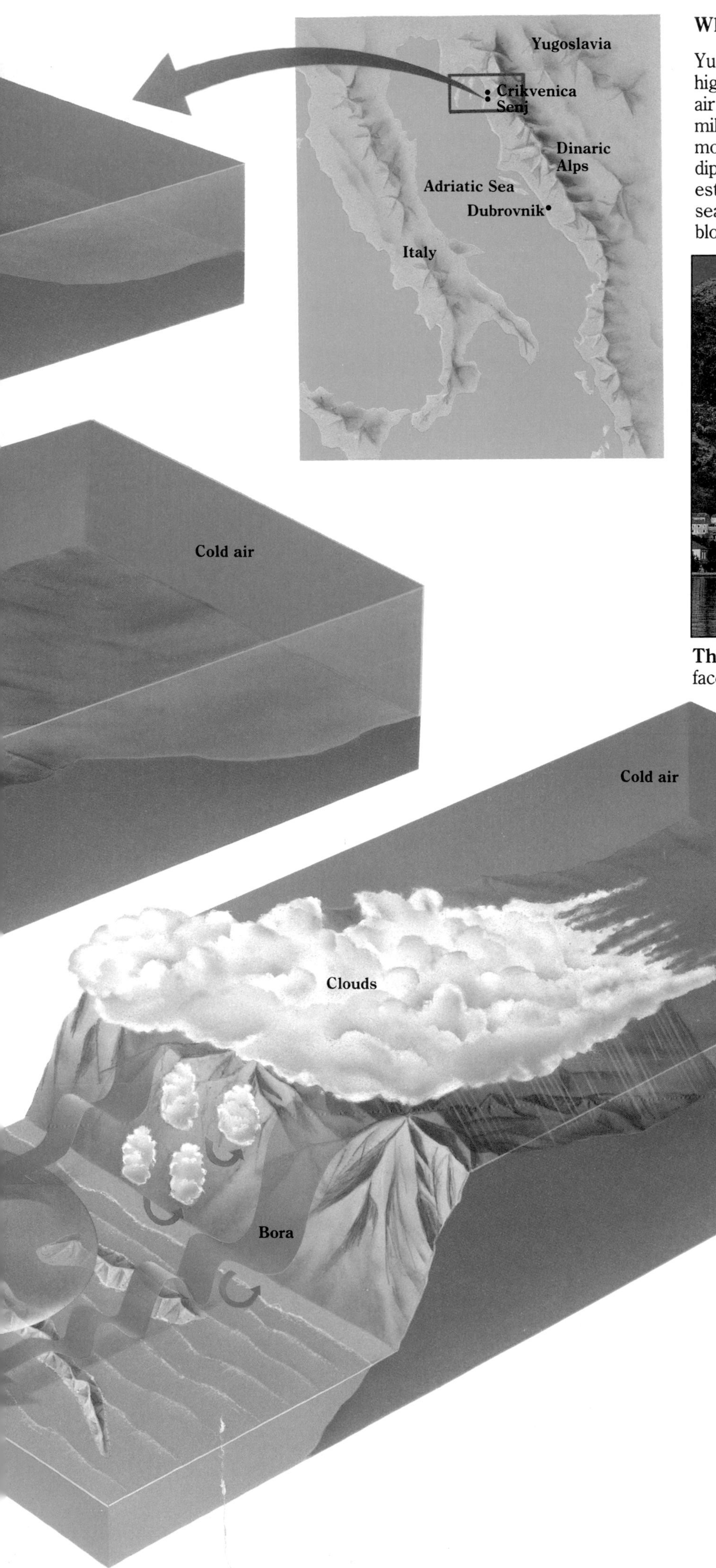


**3 The cold, dry bora** blows at enormous speeds down the mountainside. As it crashes onto the plain below, the wind curls into eddies and whirlpools. On the eastern side of the mountains, clouds form and rain falls. Sometimes the bora even brings snow and ice to this otherwise mild area of the Alps.

# How Do Tornadoes Form?

Tornadoes, the offspring of severe thunderstorms, develop out of two basic ingredients in the four stages shown on these pages. The first ingredient is supplied when a cold front runs into a mass of warm, moist air. The resulting powerful updrafts produce huge cumulonimbus clouds. As more warm air feeds into the growing clouds, downdrafts form, triggering thunderstorms.

But not every thunderstorm spawns a tornado. That requires the second ingredient—rotation. If strong cross winds blow through the cumulonimbus cloud, they can twist the updrafts into a turning mass of air. This vortex draws even more warm air into the cloud, which makes the air spin still faster. The spiral tightens, gaining speed in much the same way that an ice skater spins faster as she draws in her arms. Soon a funnel cloud, packing winds of up to 300 miles per hour, drops out of the cloud, ready to wreck houses, lift automobiles, or toss an 800-pound ice chest 3 miles.

▲ **An advancing tornado** kicks up a dust cloud.

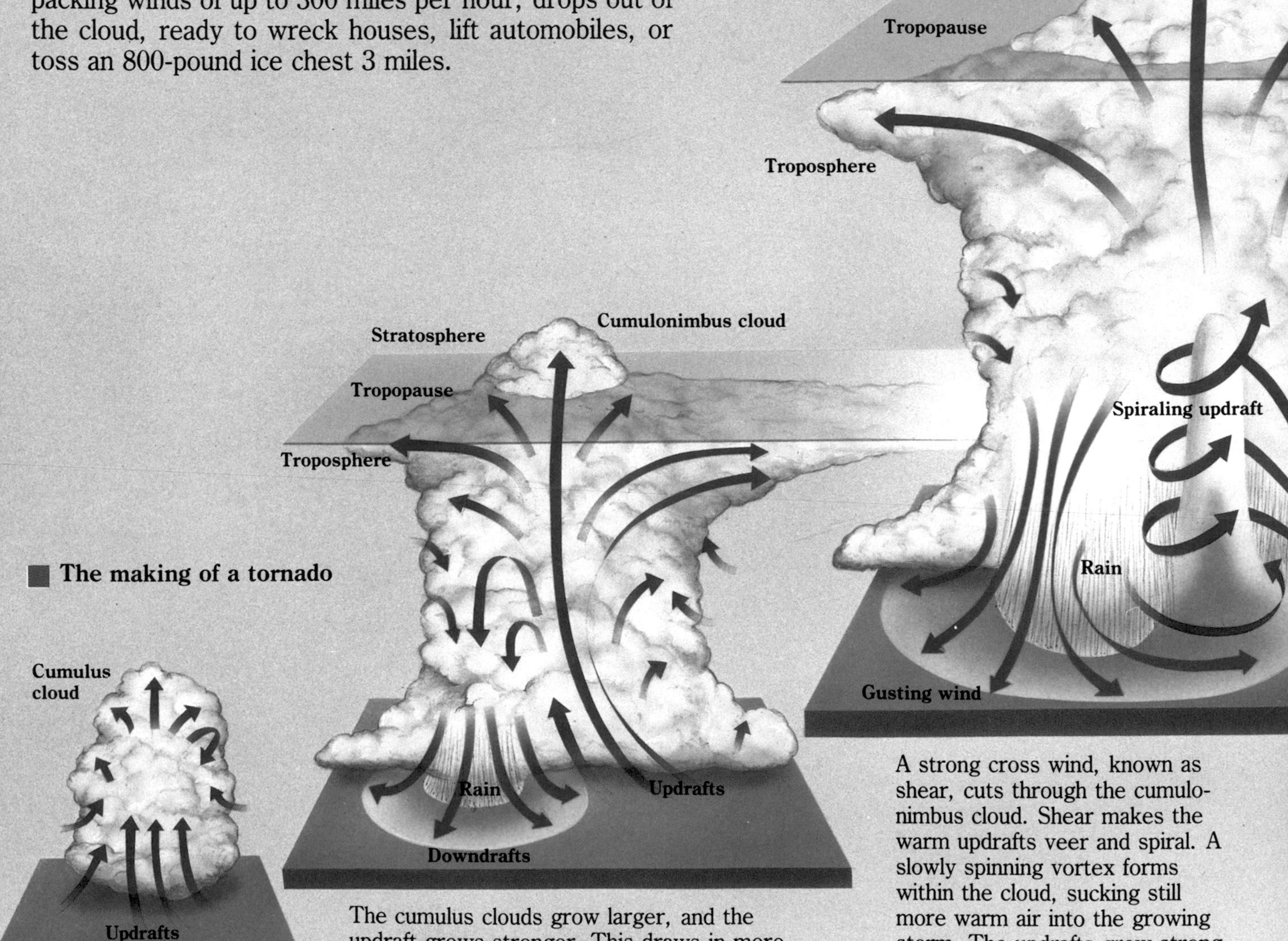


**The making of a tornado**

When cold and warm air meet, the cold air moves down. The warm updraft carries moisture into the colder upper atmosphere, and cumulus clouds form.

The cumulus clouds grow larger, and the updraft grows stronger. This draws in more warm air, turning the cumulus clouds into storm-generating cumulonimbus clouds. The tops of these towering giants reach into the cold stratosphere, so the ascending air becomes cold. This chilling creates strong downdrafts that bring rain and create a long series of thunderstorms called a squall line.

A strong cross wind, known as shear, cuts through the cumulonimbus cloud. Shear makes the warm updrafts veer and spiral. A slowly spinning vortex forms within the cloud, sucking still more warm air into the growing storm. The updrafts grow stronger, and so do the downdrafts. The spiral grows tighter and spins faster, producing a whirlpool-shaped updraft that can reach speeds of more than 65 miles per hour.

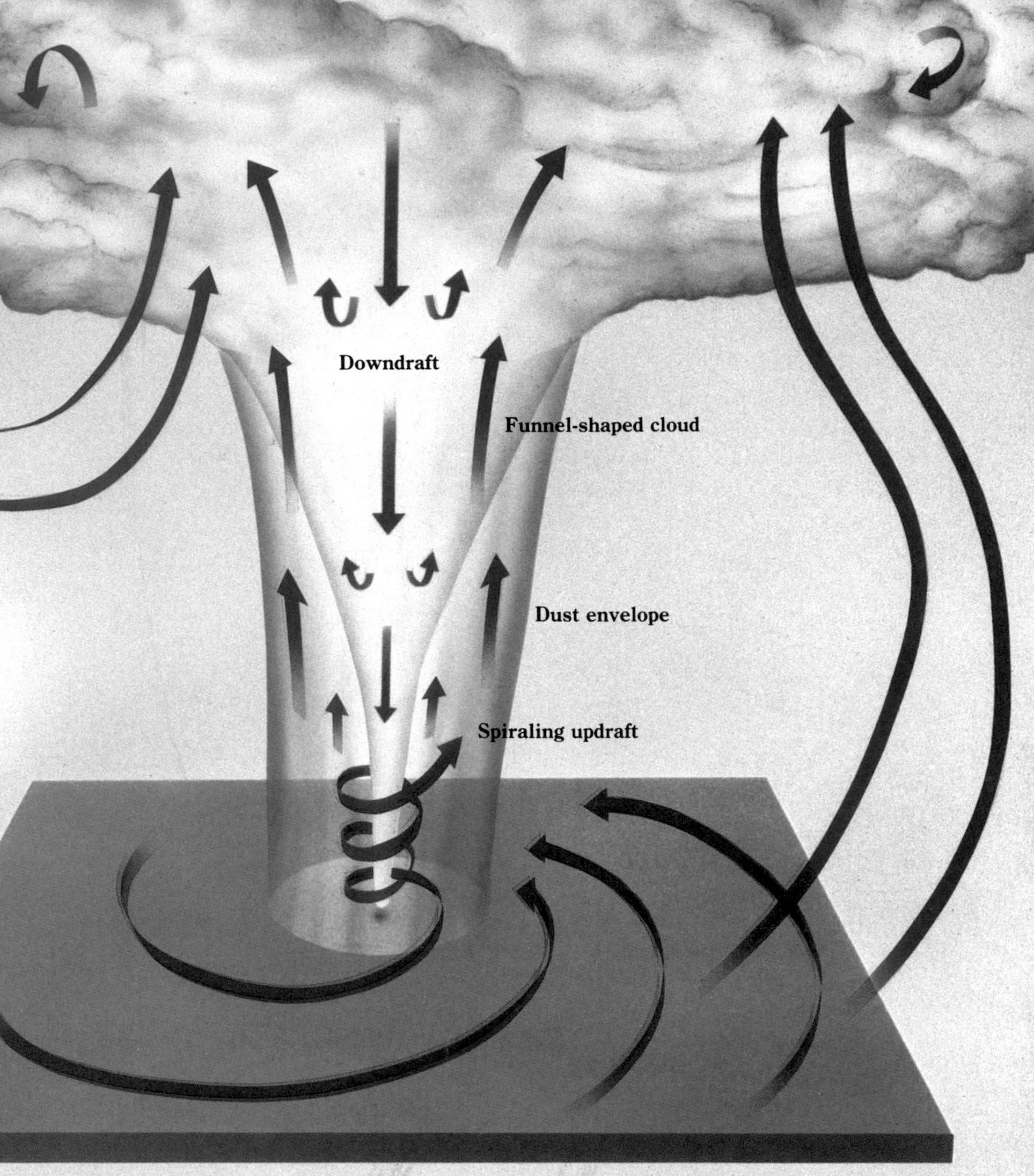


Once the whirlpool-shaped updraft forms, a spinning funnel cloud begins stretching downward. As the tornado becomes more intense, the funnel cloud grows larger and larger. Finally, the funnel touches down with explosive force. Within this powerful storm, weak downdrafts form in the area of lowest air pressure. This is the core of the storm.

## The squall line

Cumulonimbus clouds form along a cold front, producing a line of thunderstorms *(right)* called a squall line. A typical squall line is 12 to 30 miles wide, can stretch for 100 miles, and moves at a speed of 30 miles per hour. Tornadoes often develop at the southern end of a squall line.

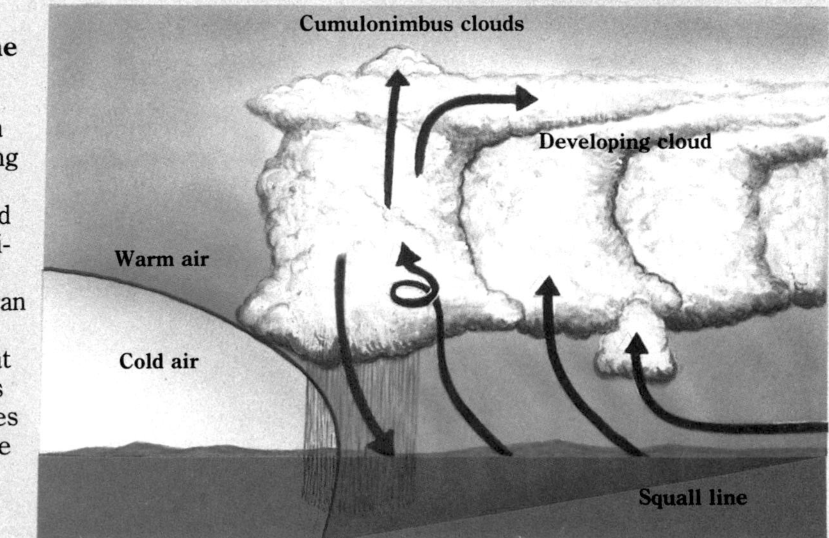

# What Causes Air Turbulence?

Most people who have flown in an airplane have experienced the nerve-rattling ups and downs caused by air turbulence. The strong updrafts, downdrafts, and eddies that create air turbulence form when the smooth flow of air is disrupted by differences in temperature or terrain.

Thermal turbulence—that caused by differences in air temperature—comes from the fact that various features of the Earth's surface radiate heat at different rates. Roads and sandy areas, for example, heat the air overhead more quickly than do forests or lakes. Updrafts form in warmer air, while downdrafts can occur in cooler air. Even where temperatures are fairly equal, mountains and tall buildings can create turbulence by blocking the smooth flow of air. Eddies and whirlpools form in air that has passed over these obstacles.

**A swirling stationary cloud** marks the boundary between cold air and rising warm air over a volcanic peak. Clouds like this reveal thermal air turbulence.

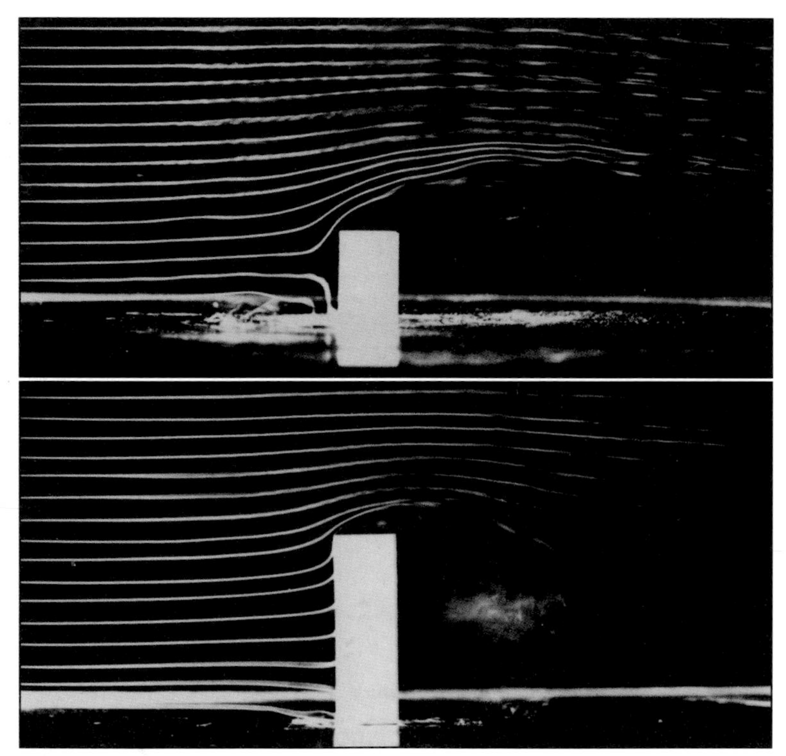

**Tall buildings create air turbulence** by altering the path of steady surface-level winds. In the photos above, several lines of blown smoke show the air-flow patterns around models of a short building *(top)* and a tall one *(bottom)*. In each case, some of the wind hitting the building is deflected upward at an angle. Some mountains have a similar effect on airflow.

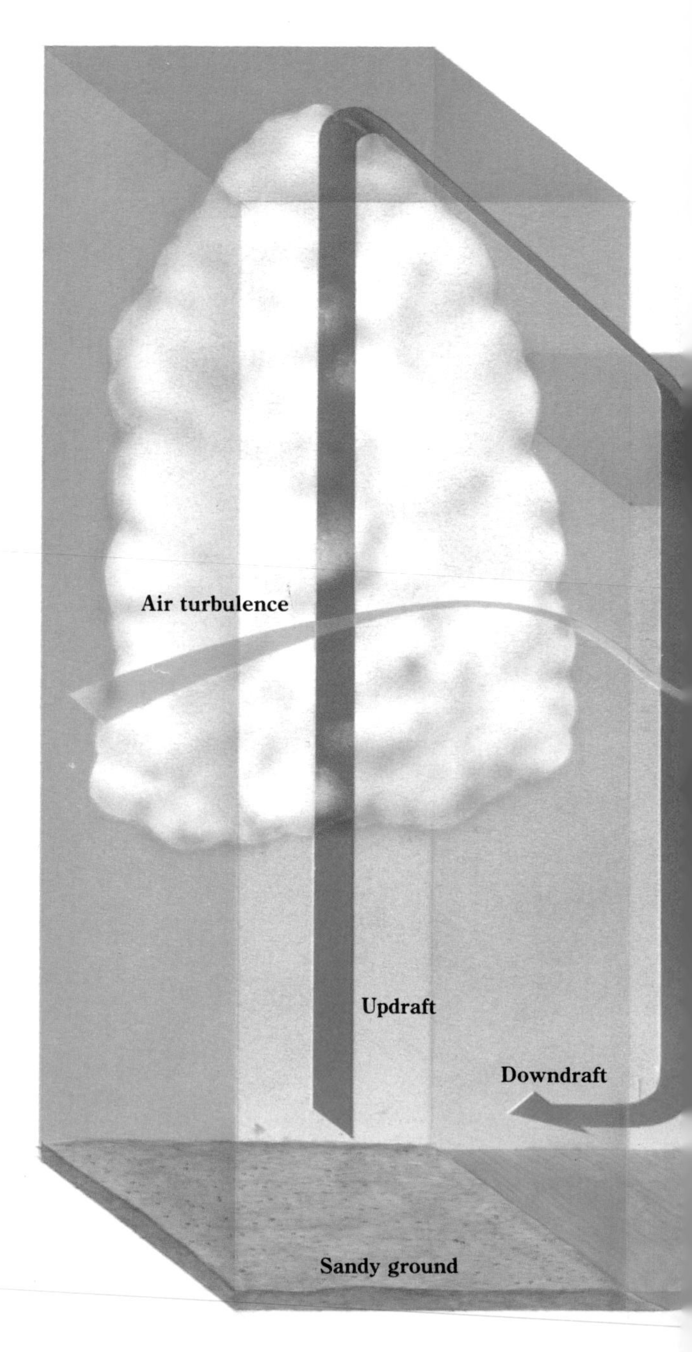


**Turbulence over varied terrain**

A jet flying over varied terrain *(right)* may be buffeted by thermal air turbulence. Strong updrafts, marked by clouds, develop over large, sandy patches of ground that warm quickly under a bright sun. Forests and rivers warm more slowly, so downdrafts sometimes form above these features of the terrain. Alternating patches of clear sky and clouds can signal a rough ride ahead.

## Turbulence in a clear sky

Stratosphere

Likely area for clear-weather turbulence

Edge of Ferrel cell

Edge of polar circulation cell

Troposphere

Jet stream

Cold air

Front

Clear sky is usually calm sky, but this is not always true in the upper atmosphere. Clear-weather turbulence often occurs near the polar front jet stream, about 6.2 miles up. Here, where the frigid polar circulation cell and the warmer Ferrel circulation cell meet, at the boundary between stratosphere and troposphere, air masses move at sharply different speeds and strong eddies can form.

## Atmospheric mountain waves

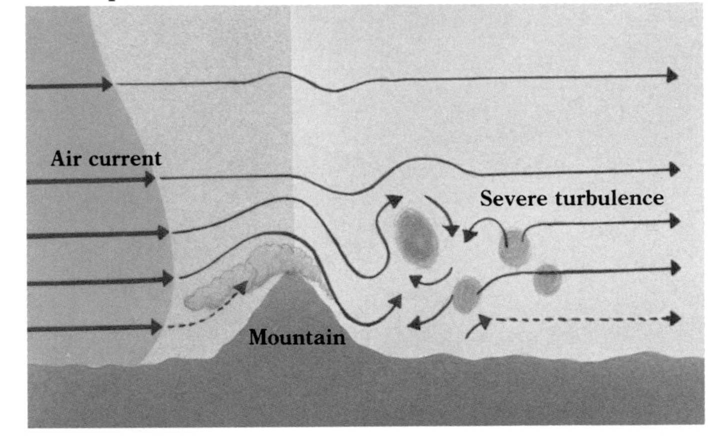


Strong mountain-wave turbulence can develop when strong winds flow over a mountain range *(above)* and upward, and then are trapped under a stable layer of air about 10,000 feet above the peaks. Deflected earthward, the winds surge down, then eddy and twist upward. Lens-shaped clouds above the range disclose the presence of mountain waves, which are common over the Rocky Mountains.

# How Does Wind Act near Buildings?

In a city with skyscrapers, the winds at street level can often take strange and unexpected paths. A tall building standing among shorter buildings presents a solitary impediment to a strong wind blowing through the city. The wind cannot pass through the obstacle, so it takes paths around and over it, splitting into downdrafts and slanting air currents.

In addition, winds deflected by several nearby buildings can converge into gusts that sweep people off their feet. Some parts of Chicago's Michigan Avenue are fitted with handrails to help pedestrians keep their footing when strong building winds blow.

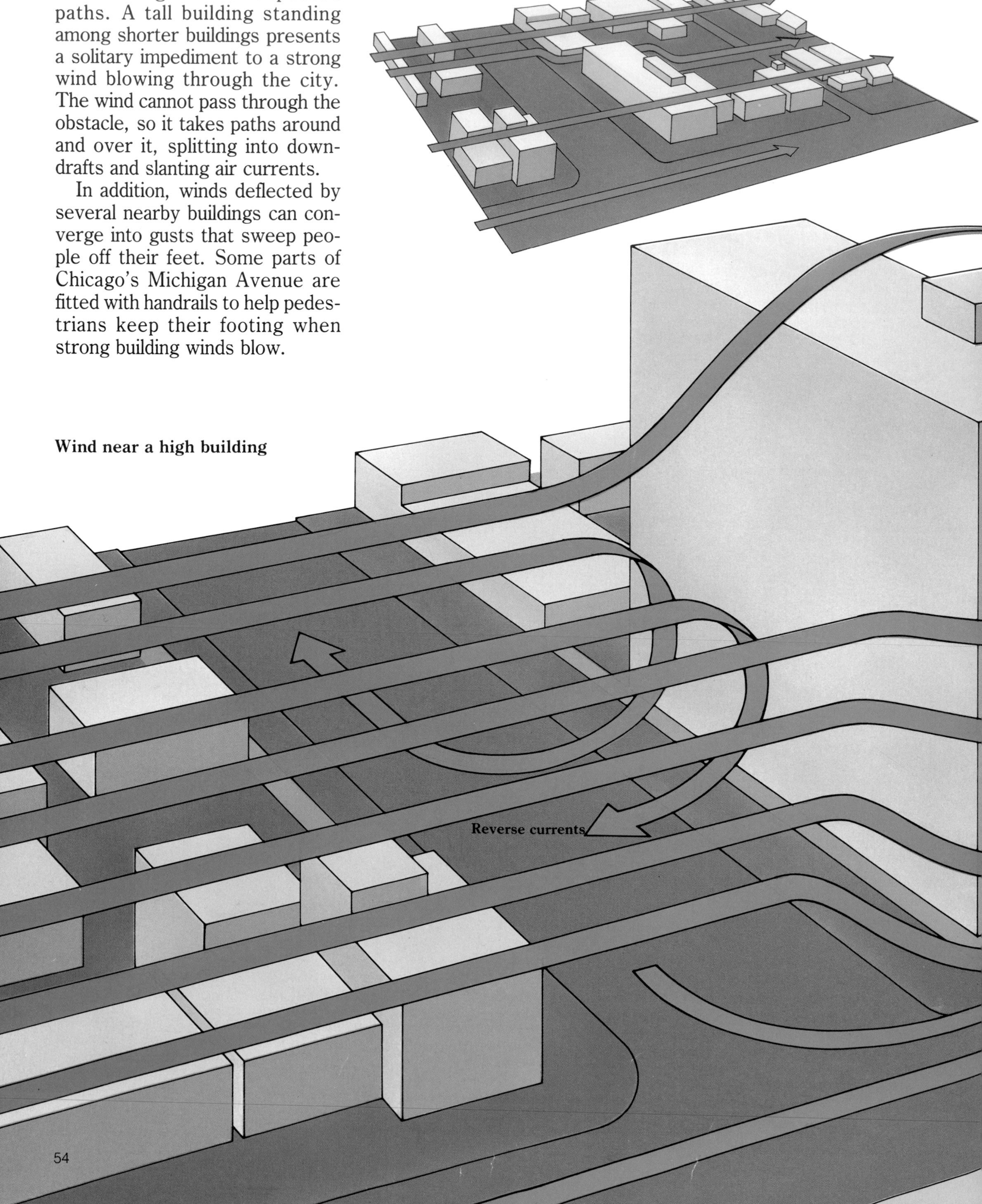

## A variety of building winds

Wind striking a tall building divides into several air flows. Some air flows down the building's face, hits the pavement, and becomes a reverse wind. Air also flows left and right, wrapping around the building's corners and veering down toward the street *(near right, top)*. Air flowing along the sides of the building caroms off the corners and becomes a fast, separated flow *(far right, top)*. Winds deflected from adjacent tall buildings can merge, creating powerful valley and street winds *(far right, bottom)*.

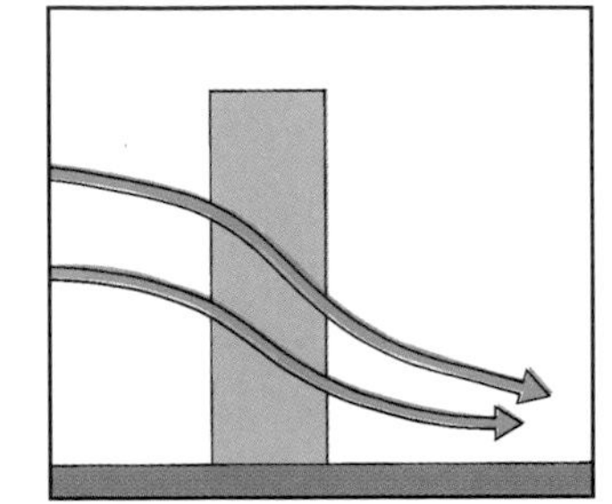

**Downward wind**

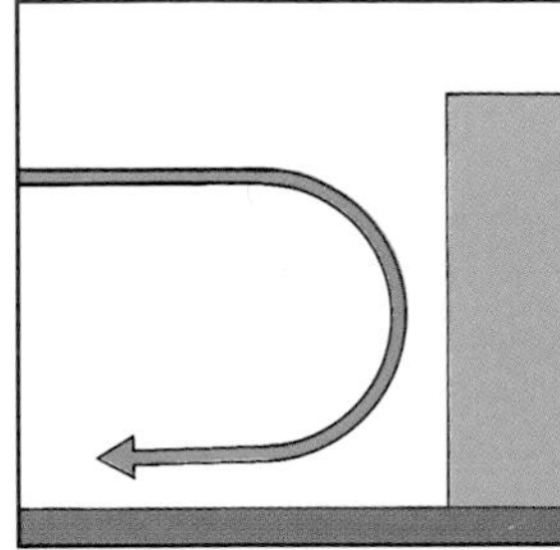

**Reverse wind**

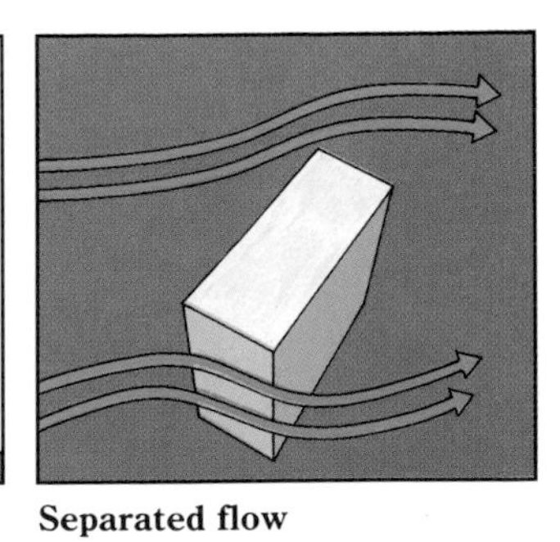

**Separated flow**

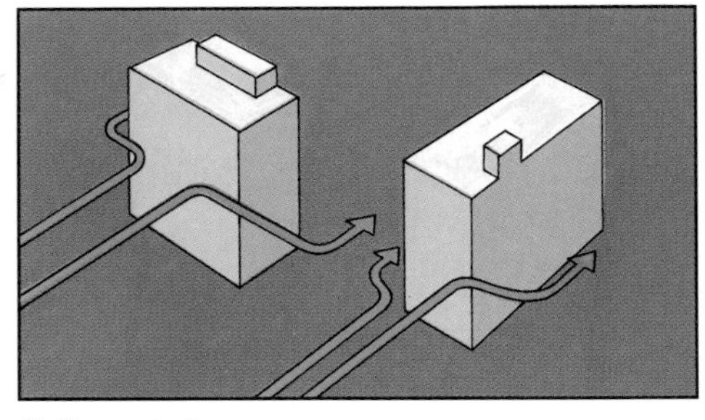

**Valley wind**

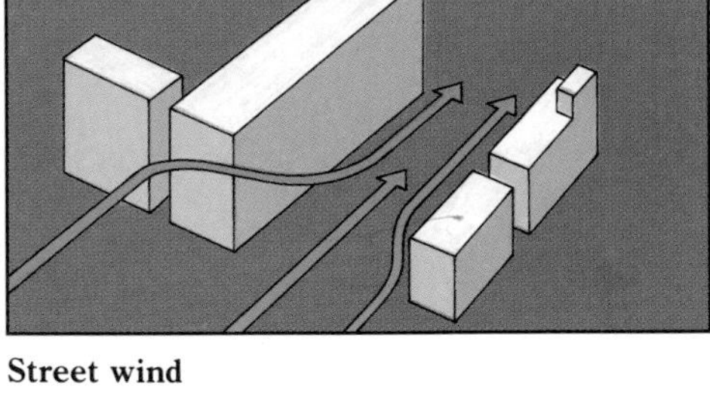

**Street wind**

Architects can test a tall building's effects on winds by placing a model of the building *(right)* in a wind tunnel. Smoke shows the air currents.

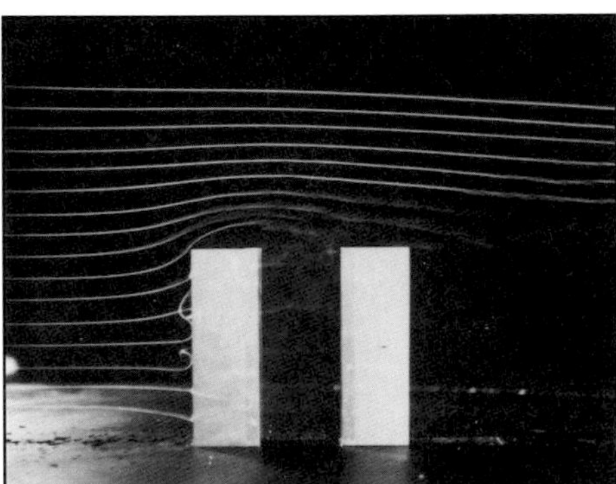

# 3
# Storm Machines

Clouds are the visible form of the life-giving water in Earth's air. Towering thunderheads, morning mist, and the delicate trails of high-flying jets are among the shapes that water takes under different atmospheric conditions. Among the substances on Earth, only water can exist in three forms—solid, liquid, and gas. And because of this, clouds are full of surprises, producing rain, snow, sleet, and combinations of precipitation throughout the changing seasons.

Within their placid exteriors, clouds—especially the cumulonimbus clouds of a thunderstorm—can be packed with action. Air currents rocket upward and drop suddenly. Water vapor condenses into a torrent of rain, freezes into snow, builds into pellets of hail, and then melts again. Electric currents build until they cannot be contained and then explode in bolts of lightning hotter than the sun's surface.

The chapter to come examines the many forms that clouds take—in the sky, at ground level, on mountain peaks—and the various ways in which they produce the rain, sleet, snow, and lightning that continually pummel the Earth.

The enormous speed of a lightning strike over a city is captured in a remarkable multiple-exposure photograph *(right)*. In this single photo, the photographer caught several flashes that occurred split seconds apart.

# How Are Rain Clouds Created?

The atmosphere can carry a certain amount of baggage in the form of water vapor that has risen from lakes, oceans, and other sources. There is, however, a limit to how much vapor the air can contain. When the air reaches that limit, which is called saturation, the vapor begins to condense and form tiny droplets of water. Clouds are nothing more than great numbers of water droplets and ice crystals that float together in the air.

Air temperature greatly influences saturation. Warm air holds more water vapor than cold air; as the temperature drops, the air becomes more humid, until it reaches the dew point, where the vapor condenses. When this happens above the ground, clouds form. When the condensed drops are large and heavy, they fall as precipitation.

**Producing a test-tube cloud**

When the plunger of the syringe is suddenly pulled back, the air inside this flask *(right)* expands and cools. This is known as insulation cooling. As the temperature falls, the water vapor in the air cools to the saturation point and forms the droplets that make up the test-tube cloud.

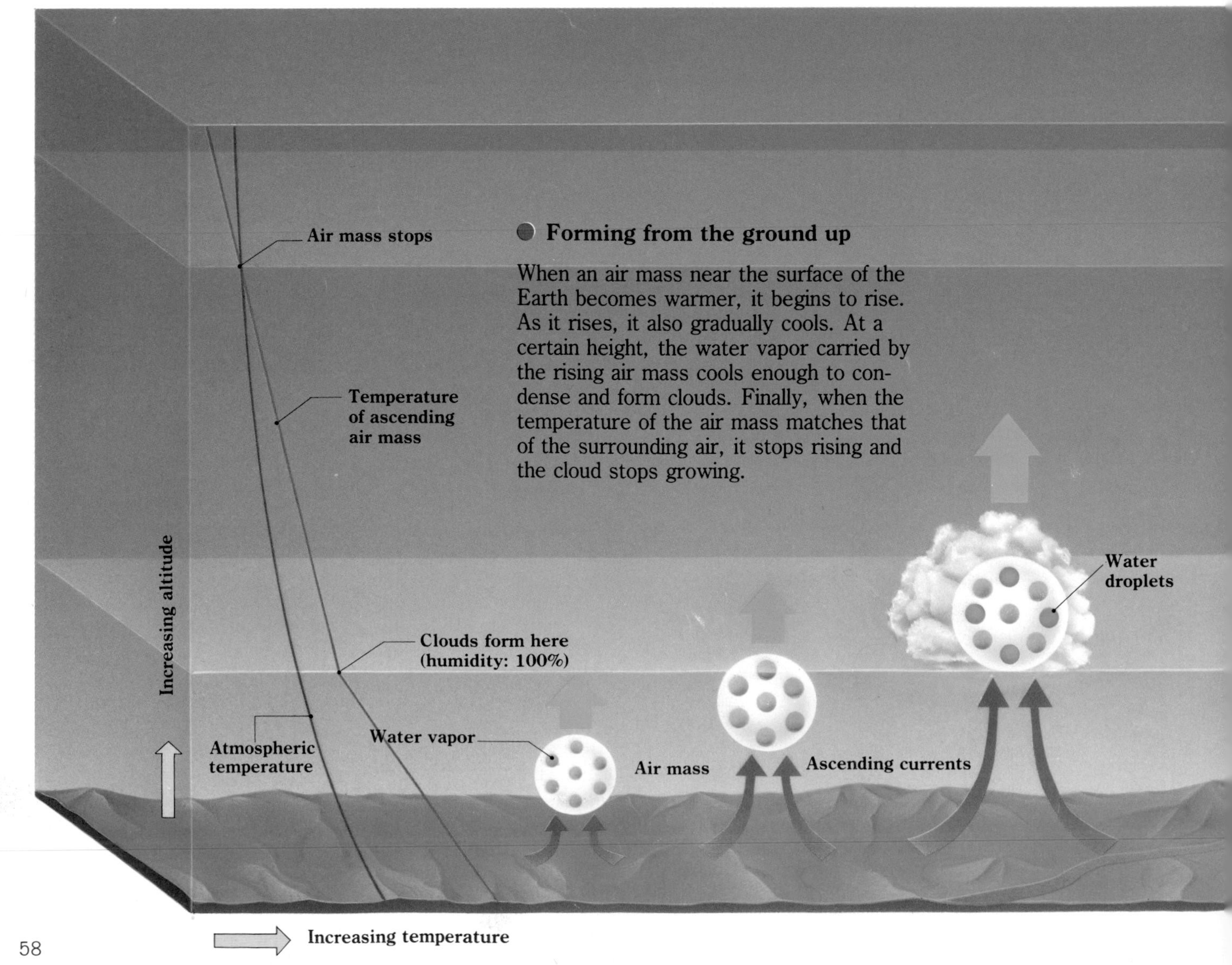


**Forming from the ground up**

When an air mass near the surface of the Earth becomes warmer, it begins to rise. As it rises, it also gradually cools. At a certain height, the water vapor carried by the rising air mass cools enough to condense and form clouds. Finally, when the temperature of the air mass matches that of the surrounding air, it stops rising and the cloud stops growing.

## Four currents that form clouds

Before clouds can form, air currents must rise. Sometimes air rises by convection, which means it was heated by the ground. Ascending currents may also occur when light warm air meets and rises over heavier cold air. An air current will also ride up and over a mountain. In addition, individual air currents meet and create strong rising currents.

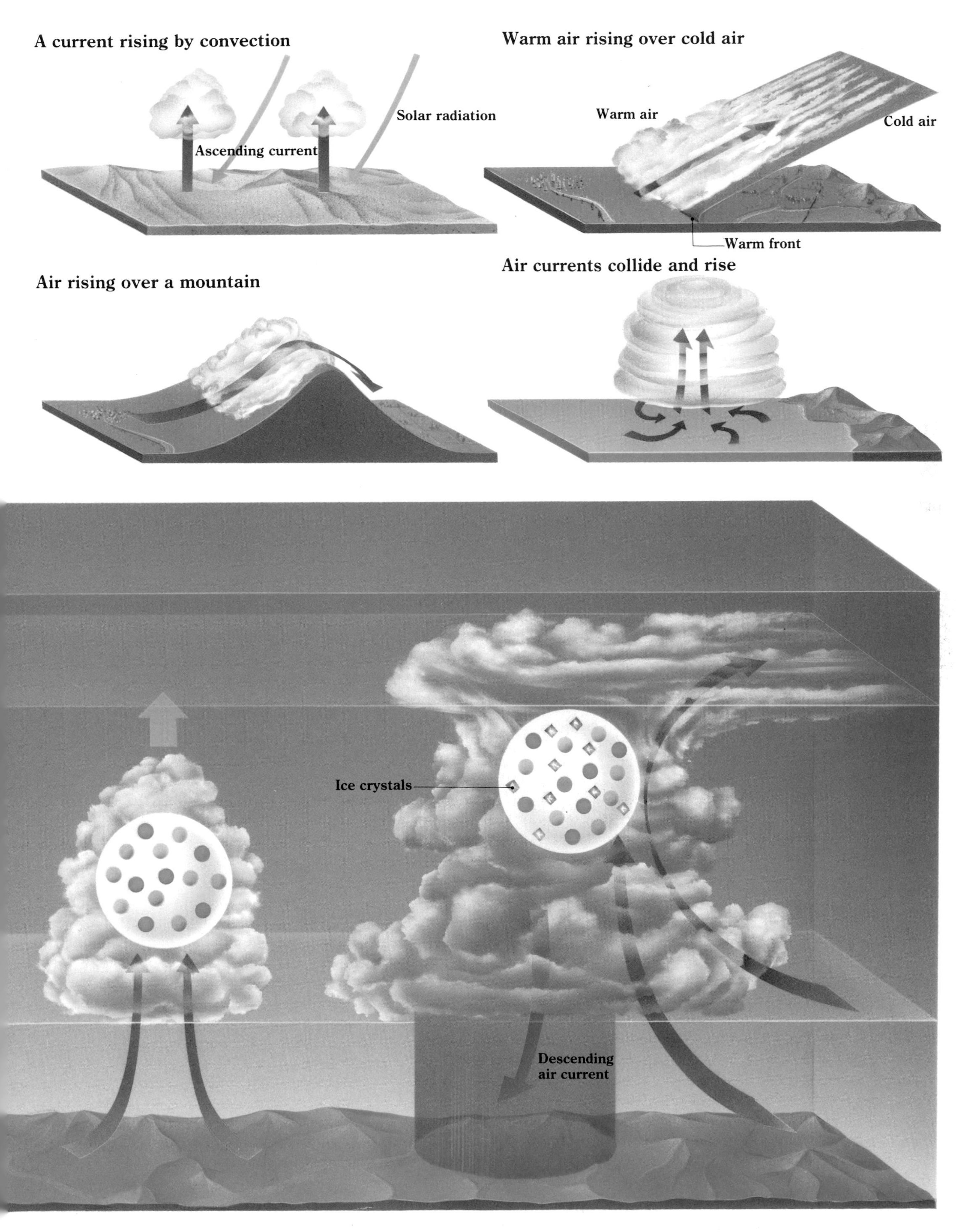

# Why Aren't Clouds the Same Shape?

Clouds generally can be classified in three main groups: layered, or stratus, clouds; clumplike, or cumulus, clouds; and striped, or cirrus, clouds. Stratus clouds develop when a large stretch of air rises slowly along the surface of a warm front. Cumulus clouds form when warm air rises from the ground or when cold air creates unstable conditions in the upper layers of the atmosphere. Cirrus clouds, on the other hand, occur when ice crystals formed in the atmosphere's upper layers fall and are spread out along prevailing air currents. These three basic types often combine to produce a great variety of additional cloud shapes.

**Cumulus clouds** *(left)* grow slowly as long as air currents continue to rise. If their growth continues they may become cumulonimbus clouds.

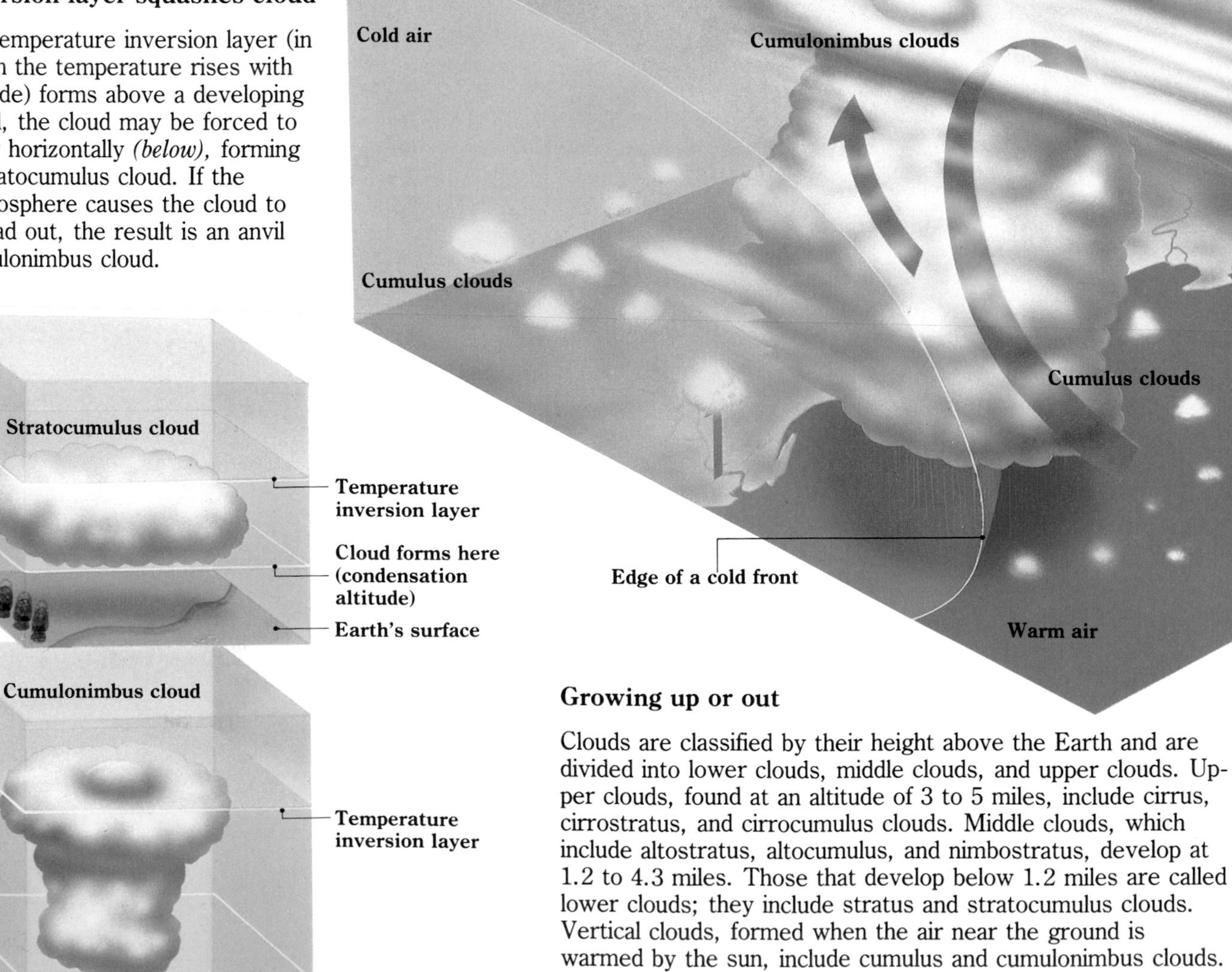


### Inversion layer squashes cloud

If a temperature inversion layer (in which the temperature rises with altitude) forms above a developing cloud, the cloud may be forced to grow horizontally *(below)*, forming a stratocumulus cloud. If the stratosphere causes the cloud to spread out, the result is an anvil cumulonimbus cloud.

### Growing up or out

Clouds are classified by their height above the Earth and are divided into lower clouds, middle clouds, and upper clouds. Upper clouds, found at an altitude of 3 to 5 miles, include cirrus, cirrostratus, and cirrocumulus clouds. Middle clouds, which include altostratus, altocumulus, and nimbostratus, develop at 1.2 to 4.3 miles. Those that develop below 1.2 miles are called lower clouds; they include stratus and stratocumulus clouds. Vertical clouds, formed when the air near the ground is warmed by the sun, include cumulus and cumulonimbus clouds.

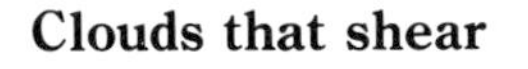

## Clouds that shear

Ice crystals from high-drifting cirrus clouds *(right)* will fall vertically if the speed of air currents is the same at all altitudes. However, they may bend, or shear, if there is a variation in speed from one altitude to another.

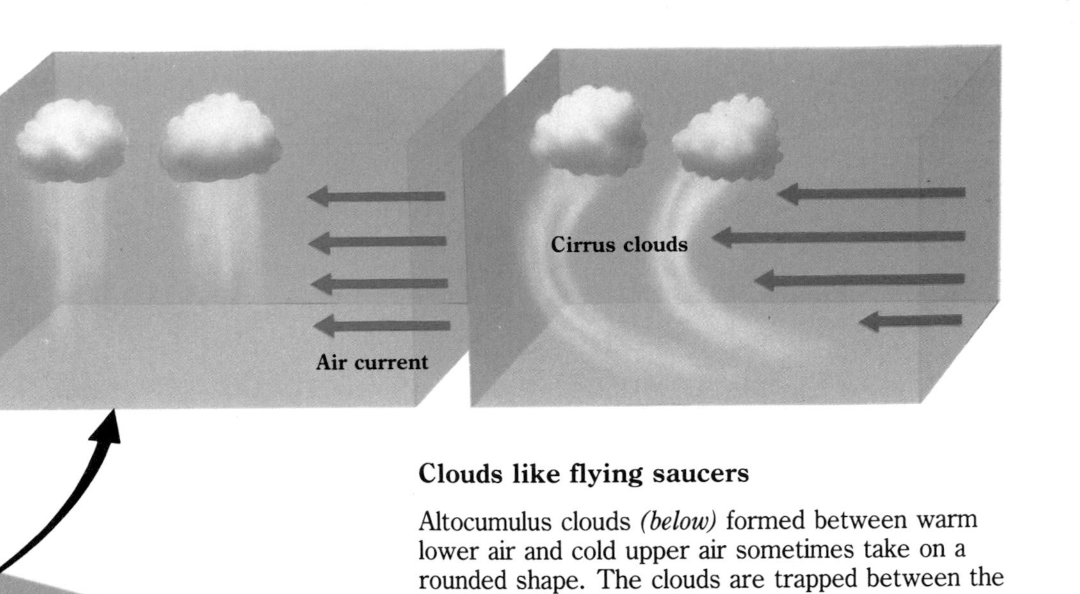

## Clouds like flying saucers

Altocumulus clouds *(below)* formed between warm lower air and cold upper air sometimes take on a rounded shape. The clouds are trapped between the falling air currents of the upper layer and the rising currents of the lower.



## Stratus clouds and rain

When rain falls on a particularly warm portion of the Earth's surface, some raindrops begin evaporating as they fall *(below)*. If the evaporation continues, the air may become saturated and form layers of stratus clouds.



## Clouds that form like waves

When horizontal air currents *(below)* are fast in the upper atmosphere and slow closer to the ground, their rolling motion produces wave-form clouds.

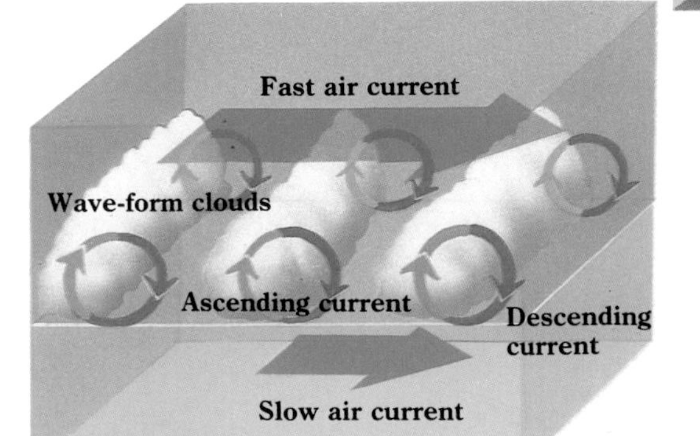


## Cresting the waves

Wave-form clouds *(right)* also develop on the crests of air currents that travel between the dry warm air above and the damp cold air below.

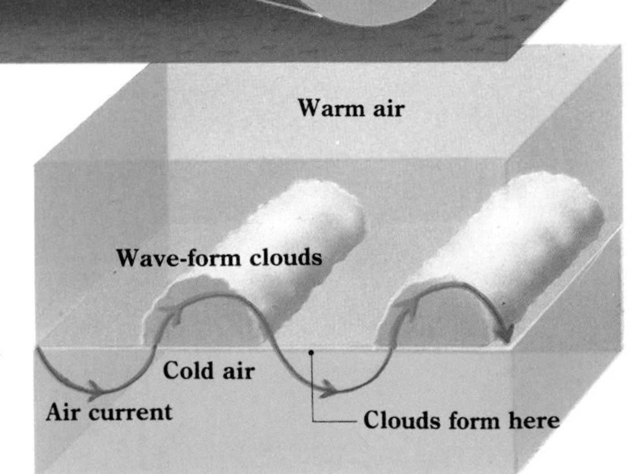

# What Are Cumulonimbus Clouds?

Cumulonimbus clouds—also known as thunderclouds because they often bring thunder and lightning *(pages 66-67)*—are the towering clouds frequently seen in midsummer. During this period, large masses of warm, moist air rise from the Earth's surface, making the atmosphere unstable. As it rises, however, the air mass begins to cool. When its temperature sinks to the dew point, the water vapor in the mass starts to condense and form a cumulus cloud.

As the cloud continues to rise and develop, it becomes a cumulonimbus cloud. Condensation increases as the cloud rises, weakening the previously strong ascending air currents and allowing descending currents to form at the bottom of the cloud. Rain begins to fall. When only the descending currents are left, the cumulonimbus clouds begin to disintegrate.

**Rising on strong ascending air currents** during a hot summer day, an anvil cloud takes shape.

**How cumulonimbus clouds form**

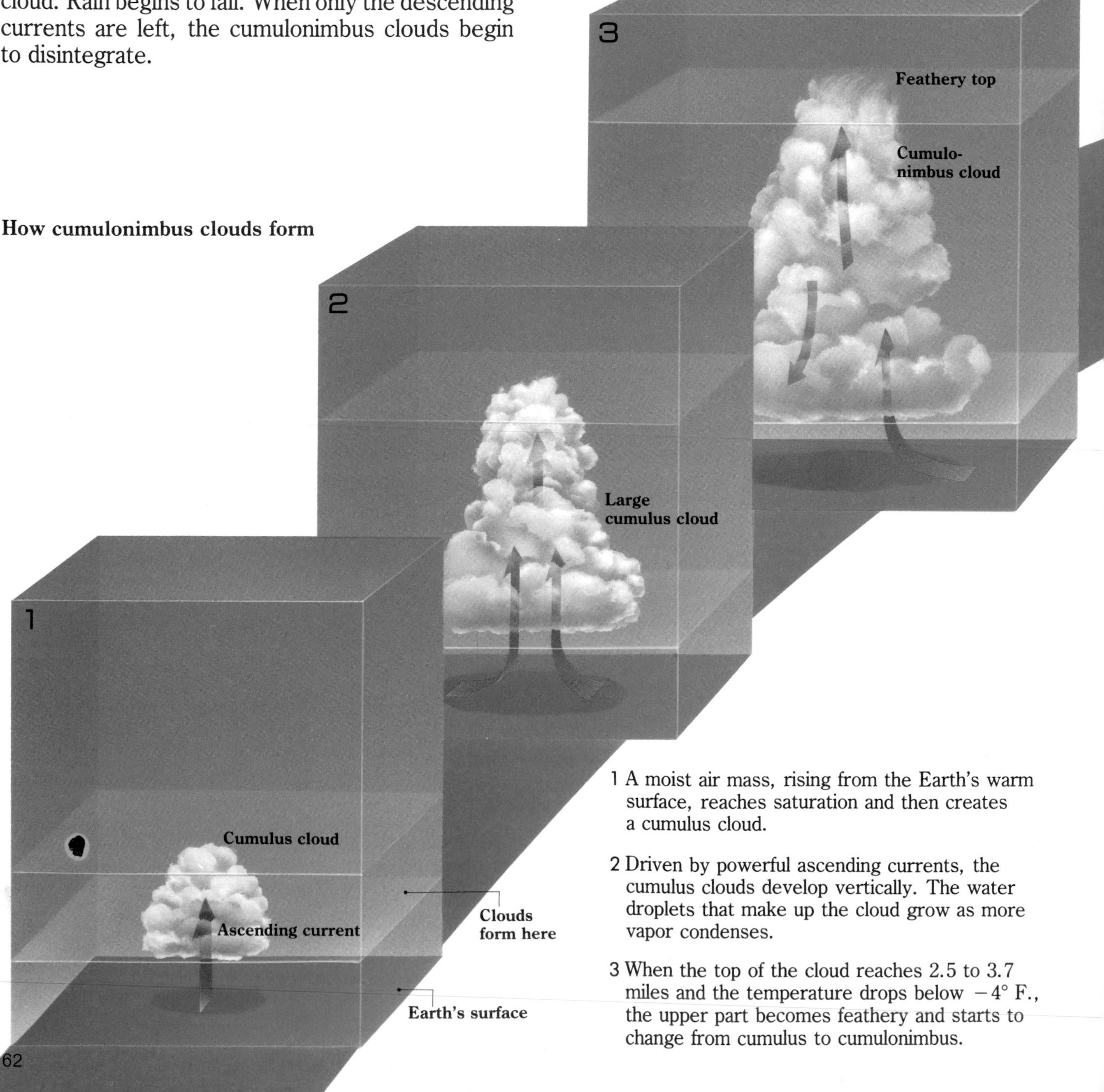


1 A moist air mass, rising from the Earth's warm surface, reaches saturation and then creates a cumulus cloud.

2 Driven by powerful ascending currents, the cumulus clouds develop vertically. The water droplets that make up the cloud grow as more vapor condenses.

3 When the top of the cloud reaches 2.5 to 3.7 miles and the temperature drops below −4° F., the upper part becomes feathery and starts to change from cumulus to cumulonimbus.

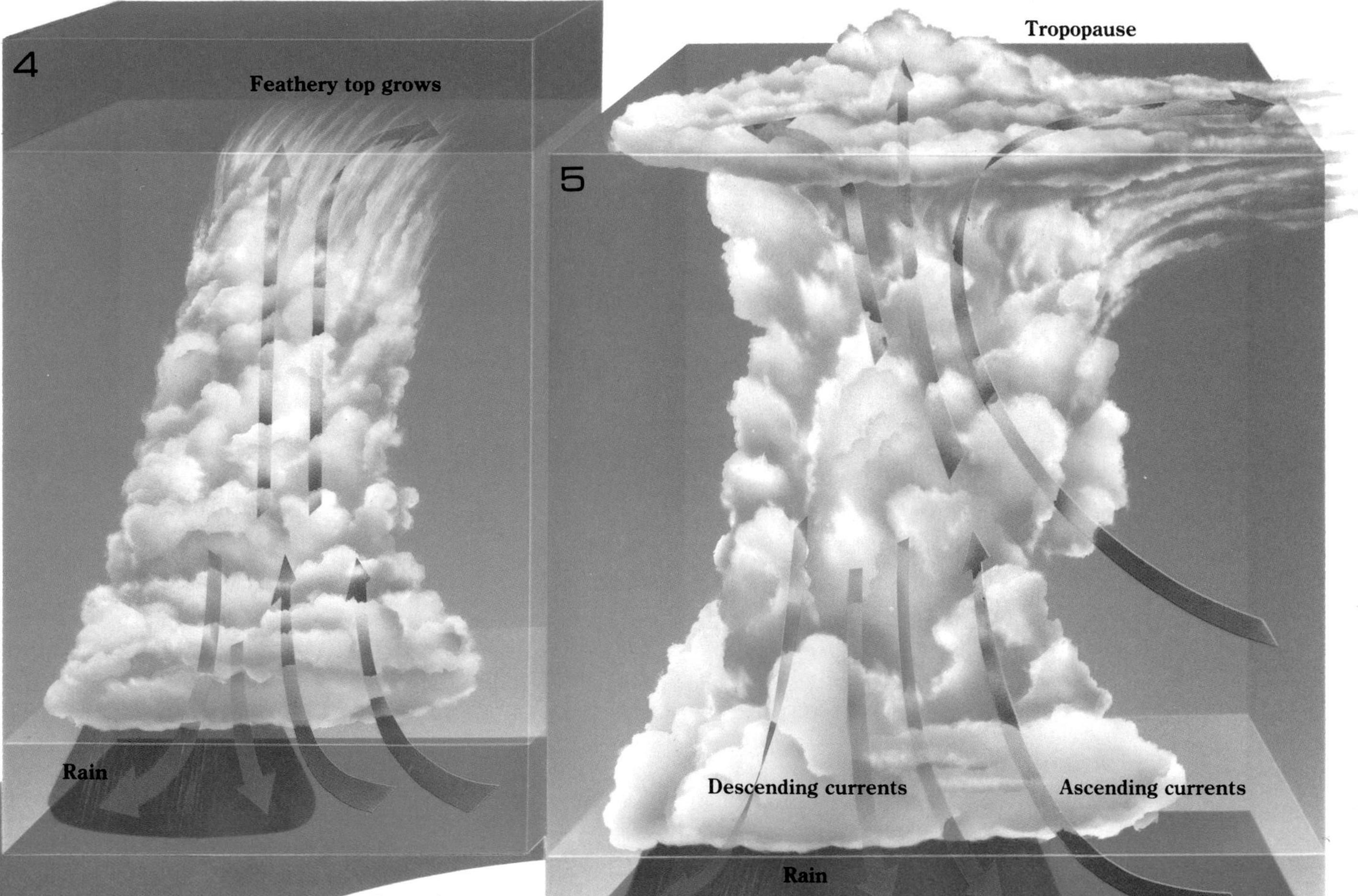


4 The cumulonimbus *(above)* develops the tall, stacked shape unique to this type of cloud. At its peak, feathery formations are visible. Rainfall increases as more water droplets form inside the cloud and descending currents become stronger.

5 The top of the cumulonimbus cloud *(above)* has reached the tropopause and will develop no further. The top of the cloud has taken on a smooth cirrus cloud form, spreading horizontally into what is referred to as an anvil cloud. At this point, the descending currents are strong and heavy rain is falling. As a result, the air is becoming cooler and ascending currents are beginning to weaken.

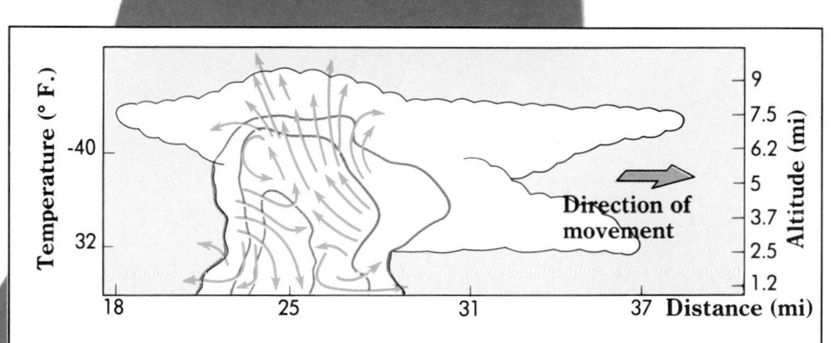


### A cloud's progress

An illustration taken from a radar image *(above)* reveals the development of a huge cumulonimbus cloud. On the right—the direction of the cloud's advance—ascending currents reach speeds of more than 99 feet per second. Descending currents on the left reach 50 feet per second.

6
Cirrus clouds
Altocumulus clouds

6 The cumulonimbus cloud *(left)* gradually breaks up when its ascending currents disappear. The silky cirrus clouds formed from the upper part of the cloud linger.

# Why Do Jets Leave Vapor Trails?

The condensation trail, or contrail, left behind as a jet streaks across the sky is really just another form of cloud. Jet-propelled airplanes, 7 miles above ground, speed through a frigid environment that encourages the formation of the ice crystals that make up cirrus and cirrostratus clouds. When a jet flies through air that is humid and cold (−22° F. or colder) at that altitude, hot water vapor from its exhaust cools quickly, providing the ice particles that form the cloudlike vapor trail.

**A contrail high in the atmosphere** begins to break up into smaller clouds.

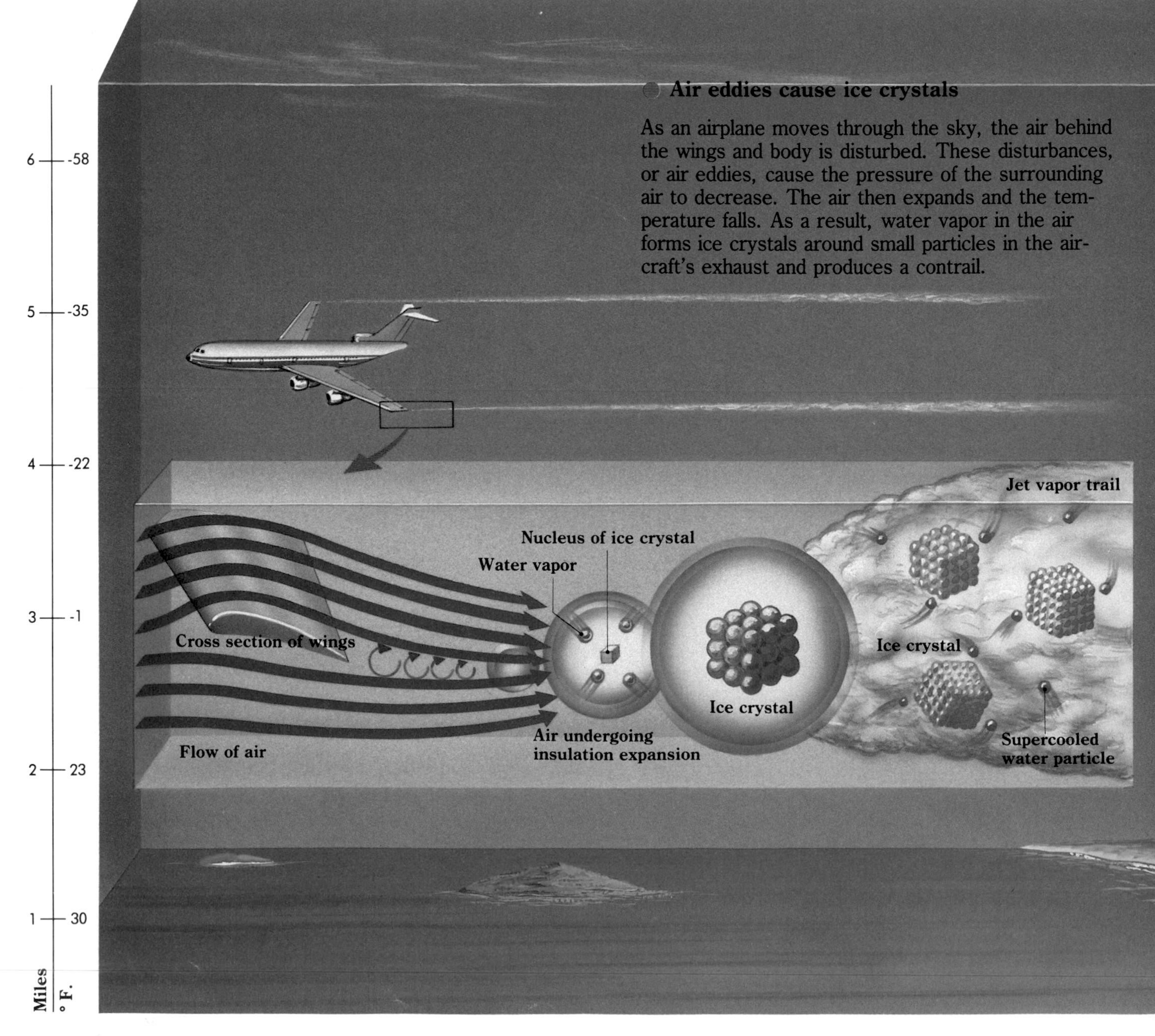
Air eddies cause ice crystals
As an airplane moves through the sky, the air behind the wings and body is disturbed. These disturbances, or air eddies, cause the pressure of the surrounding air to decrease. The air then expands and the temperature falls. As a result, water vapor in the air forms ice crystals around small particles in the aircraft's exhaust and produces a contrail.
6 -58
5 -35
4 -22
3 -1
2 23
1 30
Miles
° F.
Jet vapor trail
Nucleus of ice crystal
Water vapor
Cross section of wings
Ice crystal
Ice crystal
Flow of air
Air undergoing insulation expansion
Supercooled water particle

## How contrails change

Although contrails sometimes disappear almost immediately, they can last for an hour or more. How long depends on both the atmospheric conditions and the airplane's altitude. Once it breaks down, a contrail will normally spread out and look like other cirrostratus clouds.

A contrail holds its shape in moist air.

In dry air, a contrail spreads out.

Flying through a cirrostratus cloud, a jet produces a chainlike cloud.

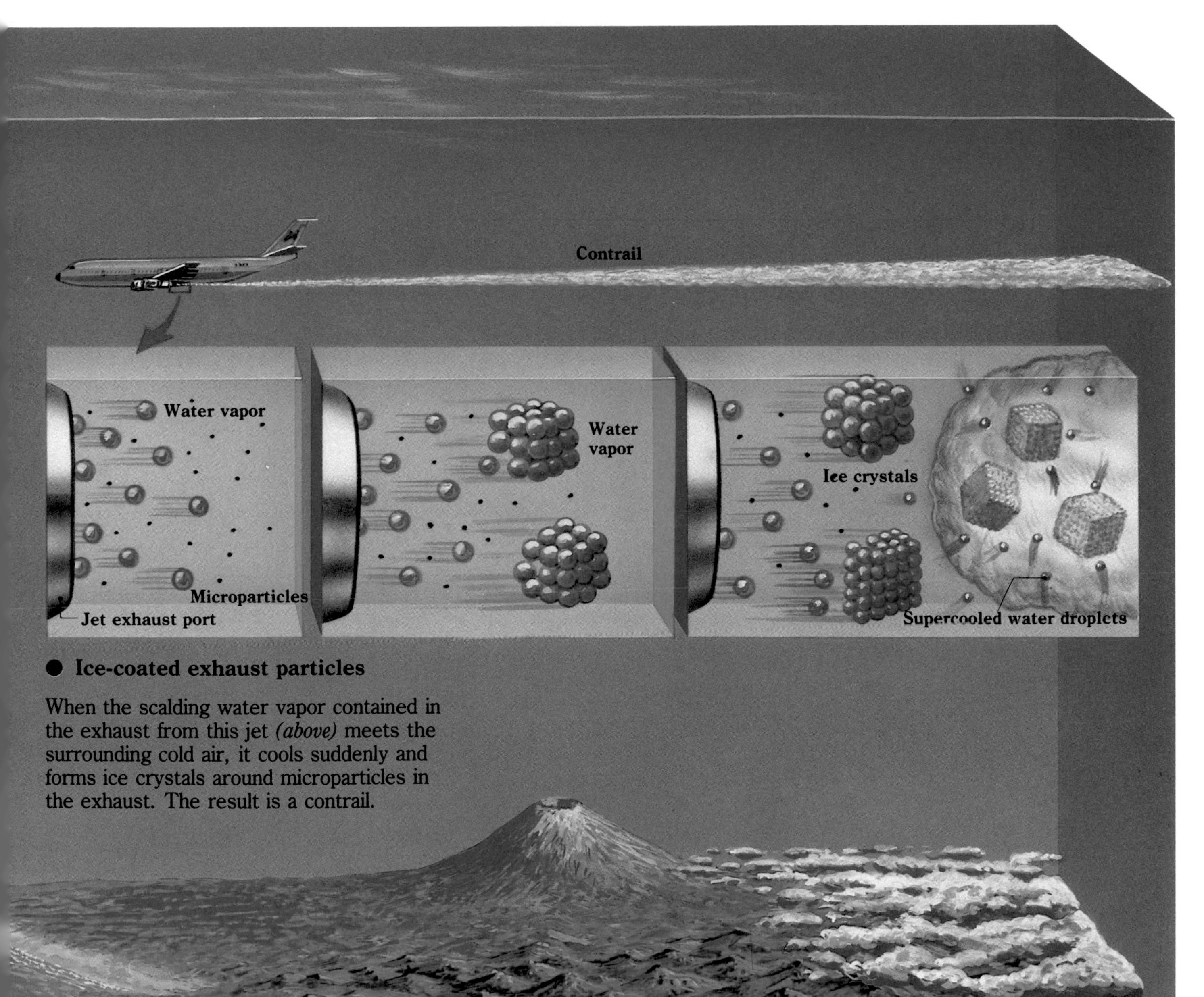


### ● Ice-coated exhaust particles

When the scalding water vapor contained in the exhaust from this jet *(above)* meets the surrounding cold air, it cools suddenly and forms ice crystals around microparticles in the exhaust. The result is a contrail.

# What Causes Lightning?

Striking 100 times a second somewhere on the planet and packing temperatures up to 50,000° F., lightning is both a common and a fearsome occurrence. Typically, these electrical flashes are found in cumulonimbus clouds, but they may also occur in nimbostratus clouds, in snowstorms and dust storms, and even in the gases of an active volcano.

Thunderstorms occur when a cloud becomes electrically charged. This can happen when ice crystals, water droplets, and other particles collide within the rising and falling air currents in clouds, producing electricity. The atmosphere usually works as an insulator to prevent this electricity from escaping. However, when the electricity stored in the thundercloud reaches a certain level, the insulation effect breaks down and allows the instant formation of an enormous electrical current, known as lightning.

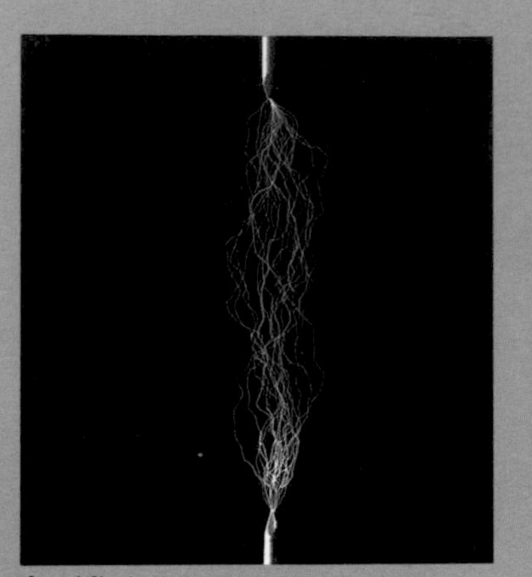

**Artificial lightning**

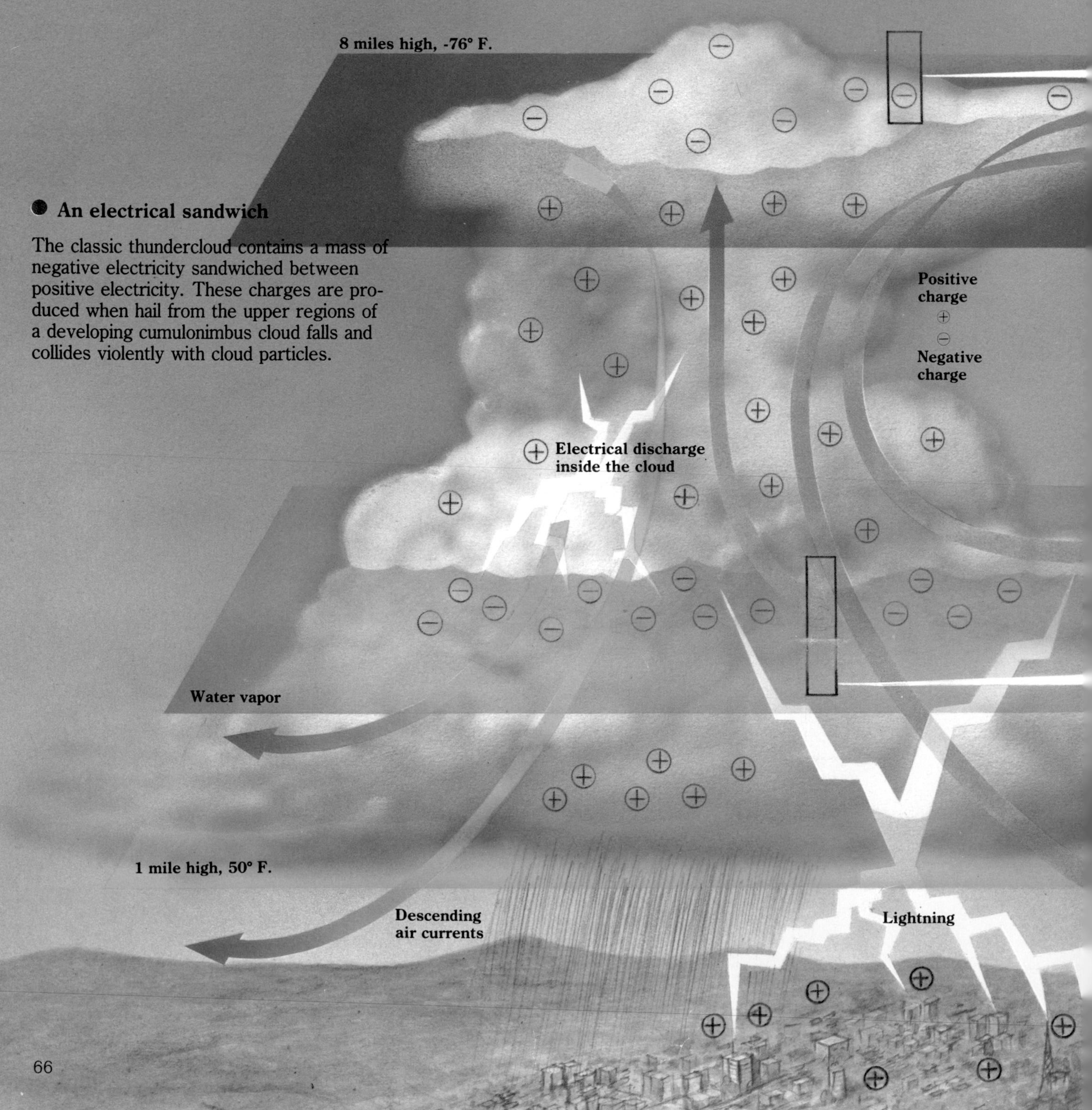


**An electrical sandwich**

The classic thundercloud contains a mass of negative electricity sandwiched between positive electricity. These charges are produced when hail from the upper regions of a developing cumulonimbus cloud falls and collides violently with cloud particles.

## Winds that build clouds

Winds moving over mountain peaks generate different kinds of clouds. If the winds approaching a mountain are extremely weak, they form a laminar (layered) flow *(figure 1)* that follows the shape of the mountain and creates no clouds. If the winds are somewhat stronger, they will generate small whirlwinds called eddies *(figure 2)* on the other (leeward) side of the mountain, but still no clouds. If the air is very humid and the winds increase in strength, wave-form currents create a lenticular (lens-shaped) cloud on the leeward side *(figure 3)* and a cap cloud near the summit. Should the winds become extremely strong, the currents on the leeward side will create eddies that move in opposite directions *(figures 3 and 4)*. If the air is very humid, the cap cloud will be accompanied by smaller clouds near the eddies.

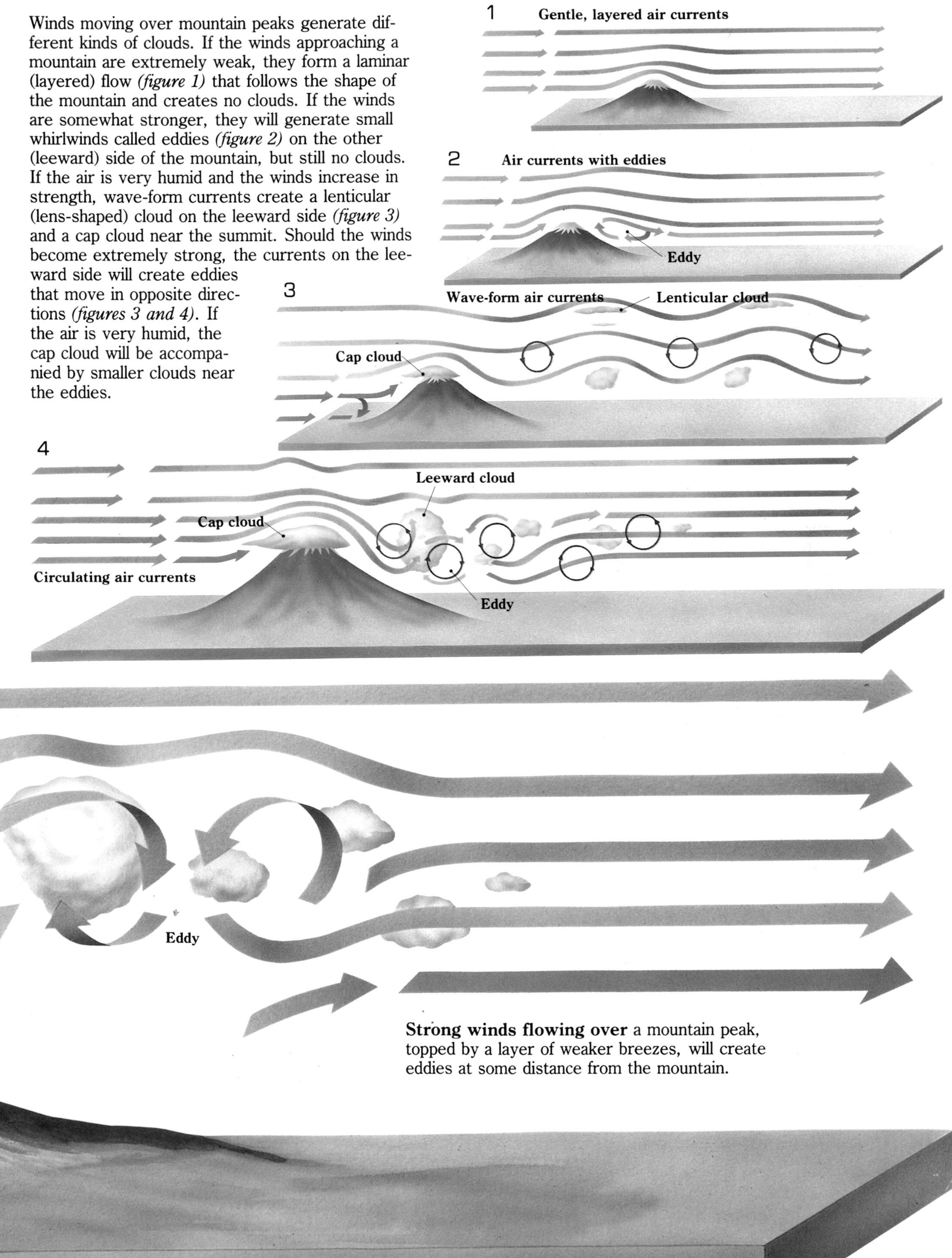


**Strong winds flowing over** a mountain peak, topped by a layer of weaker breezes, will create eddies at some distance from the mountain.

# How Does Fog Form?

Fog is actually a cloud that forms close to the ground. It appears when warm, moist air comes in contact with cooler air. Temperature determines how much water vapor the air can hold; the colder the air, the less vapor it can carry. When the air contains more vapor than it can hold at a certain temperature (a point called saturated vapor volume), the water vapor condenses into fog.

If the temperature is low enough, fog can develop even in relatively dry air. Fog also forms more easily in air that contains large amounts of dust or other particles to which the water droplets may attach themselves. In polar regions, where temperatures dip below 2° F., a frozen fog made up of ice crystals may form.

**Condensation is everywhere**

The same process of condensation that creates fog also produces some common sights of daily life. For example: Cold air on the outside of a window chills the warm air inside *(1, above)*. Water vapor condenses as the inside air cools and forms water particles, which make the window appear foggy. When the moist, warm air in a person's mouth *(2)* emerges, it cools quickly and water vapor condenses, making the exhaled breath look like fog. The cold juice inside a drinking glass *(3)* cools the air around it, making the water vapor in the air condense and form droplets on the side of the glass. Water vapor rising from a teakettle *(4)* is cooled by the air and condenses in a foglike cloud.

**How radiant fog forms**

At night, after the ground starts to release (radiate) the heat it absorbed during the day, the air near the surface begins to cool. If it cools enough, the water vapor in the air condenses into radiant fog. This type of fog often occurs in low areas on clear, nearly windless nights.

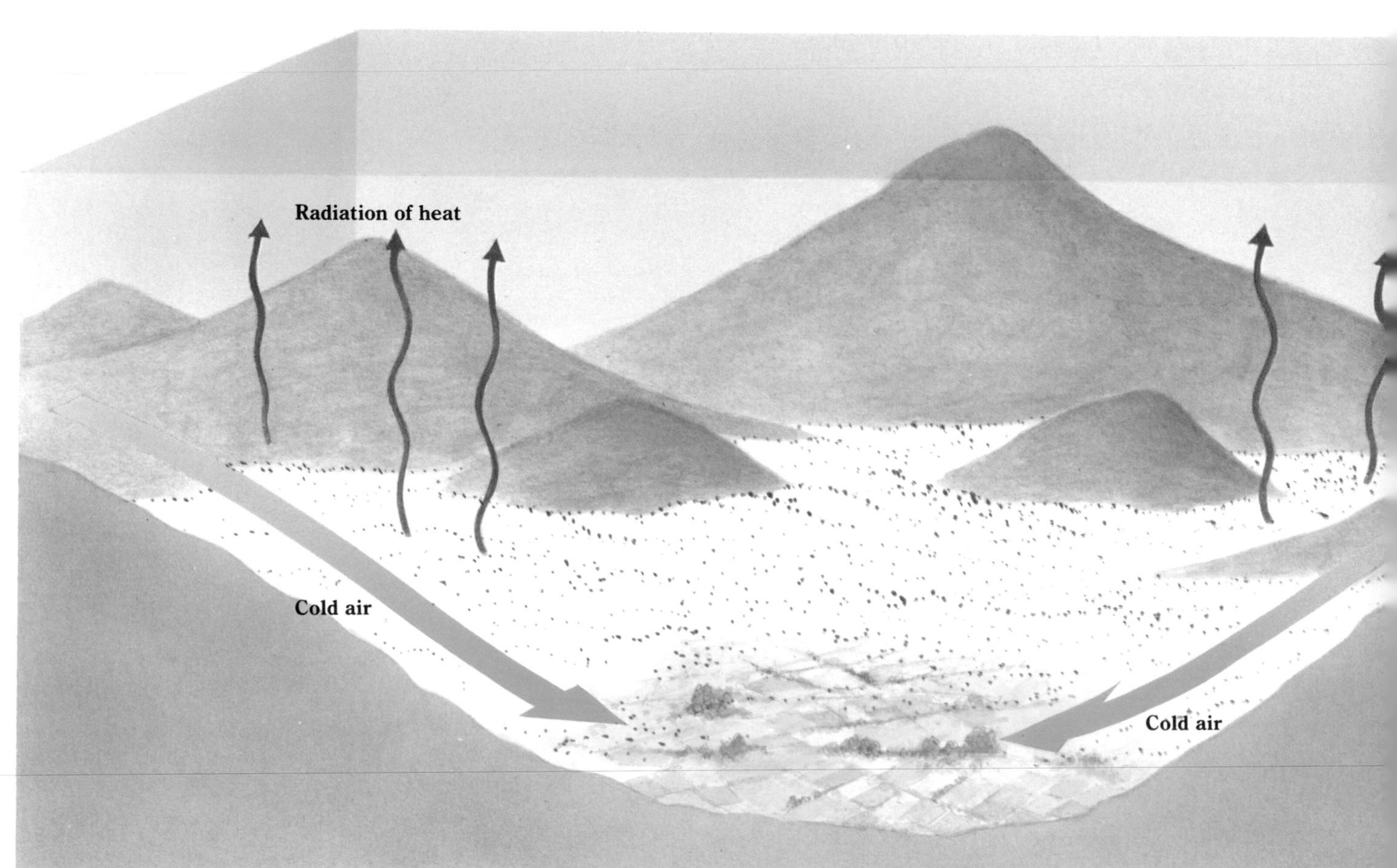

## How advection fog forms

Advection fog, the kind often found near oceans, appears when a mass of moist, warm air moves in suddenly over a cold surface. The lower layer of the air cools, causing water vapor to condense and form fog particles.

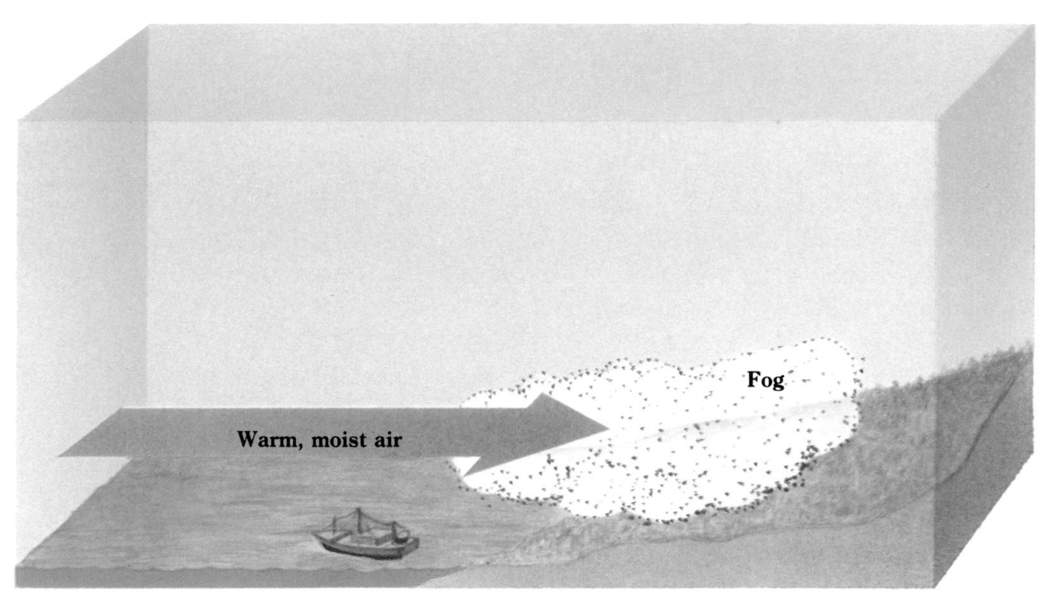


## How sliding fog forms

As warm, moist air travels up the side of a mountain, it tends to expand and cool, thereby allowing the water vapor in the air to condense into fog. This upslope fog, which may appear in irregular patches, is the type most often encountered by mountain climbers. Eventually, if the air currents continue to move upward, the sliding fog will develop into clouds.

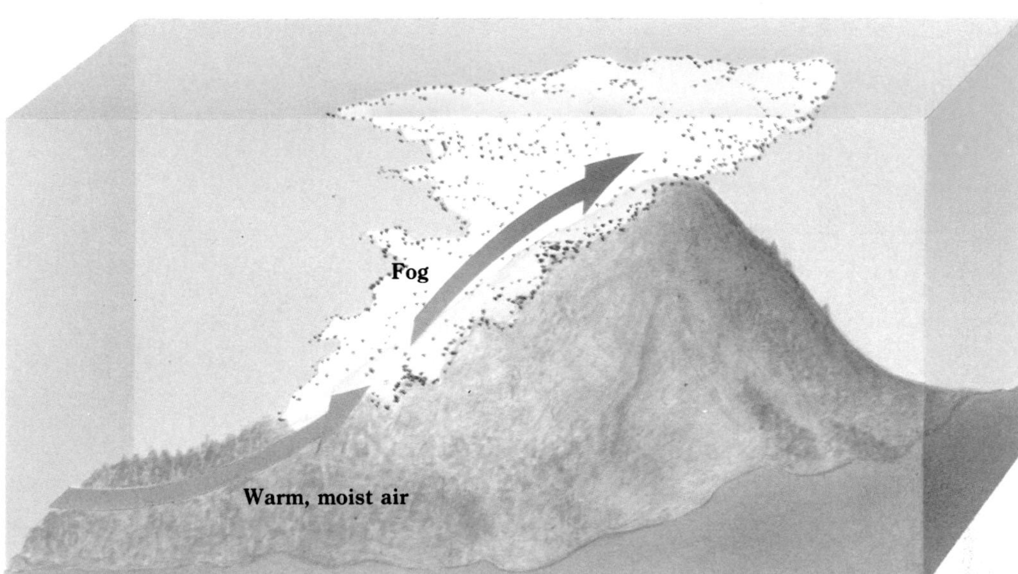

## How steam fog forms

Sometimes cold air will move over areas such as rivers or ponds that retain their heat at night. The cold air will chill and condense the warm-water vapor, forming steam fog. The greater the temperature difference between the water and the air, the thicker the fog will be.

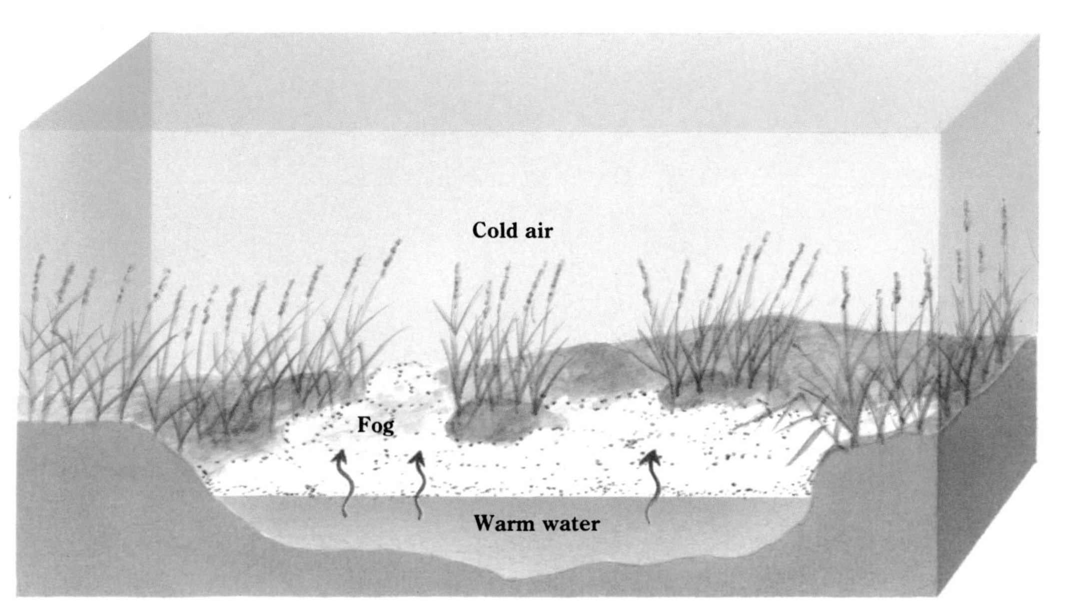

# What Is the Garúa?

Motorists driving along the coast of Peru and northern Chile sometimes experience a strange phenomenon: a fog so clear that it poses no problems with visibility, yet so wet that drivers must use their windshield wipers. This clear fog, known as the garúa, is caused by the presence of the cold Humboldt Current, which flows off the Peruvian coast. Warm Pacific Ocean air meets the cold current and forms a normal fog over the sea. However, when the fog reaches land on ocean breezes, it abruptly enters a hot, dry region, where temperatures reach 80° F. or so. As the dry air begins to evaporate the water droplets in the fog, the droplets shrink, forming unusually small particles. The result is a wet, almost invisible fog.

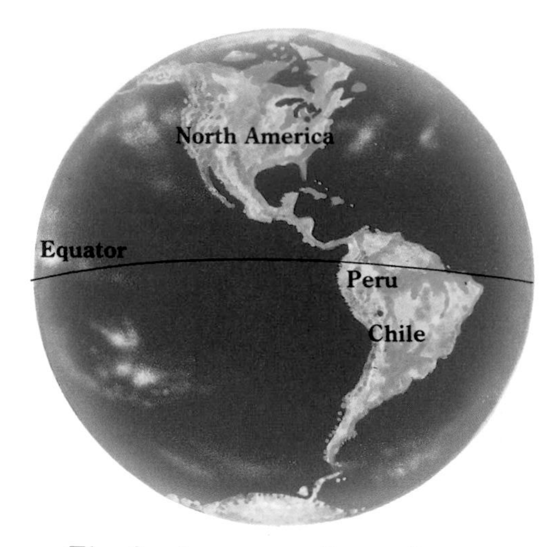


**The fog known as** the garúa occurs along the Pacific coast of Peru and northern and central Chile.

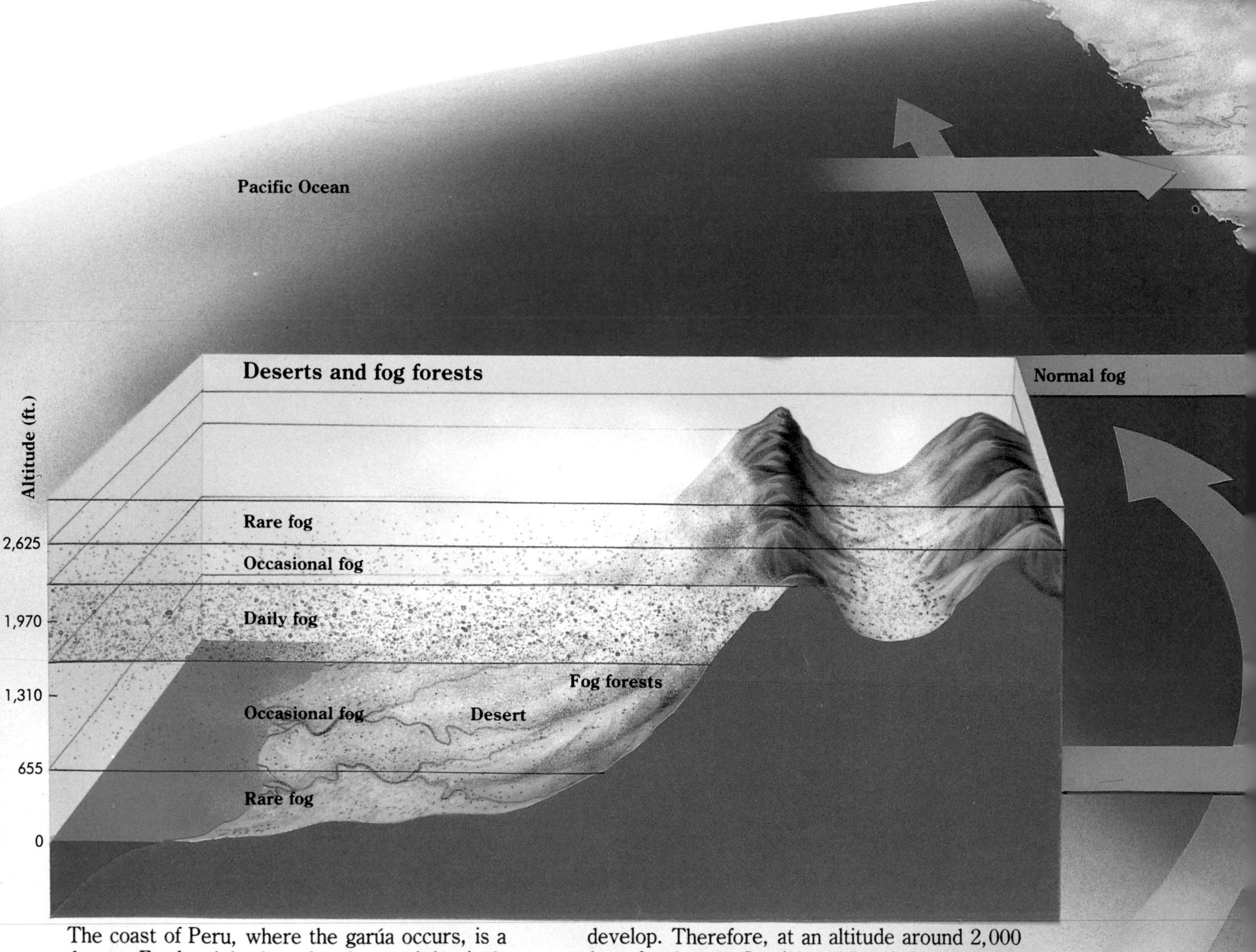


The coast of Peru, where the garúa occurs, is a desert. Farther inland on the slopes of the Andes mountains, the cooler air allows a normal fog to develop. Therefore, at an altitude around 2,000 feet, fog forests flourish, taking their moisture from the damp air.

## Rainfall in South America

As shown in the map at left, the rain that drenches some portions of South America is rare along the central Pacific coast, in some places totaling no more than 10 millimeters (mm) (.4 inch) a year. In Lima, Peru, where average temperatures are high, rainfall rarely exceeds 13 mm (.5 inch) a month *(below)*. This climate, combined with moist ocean air, creates the garúa.

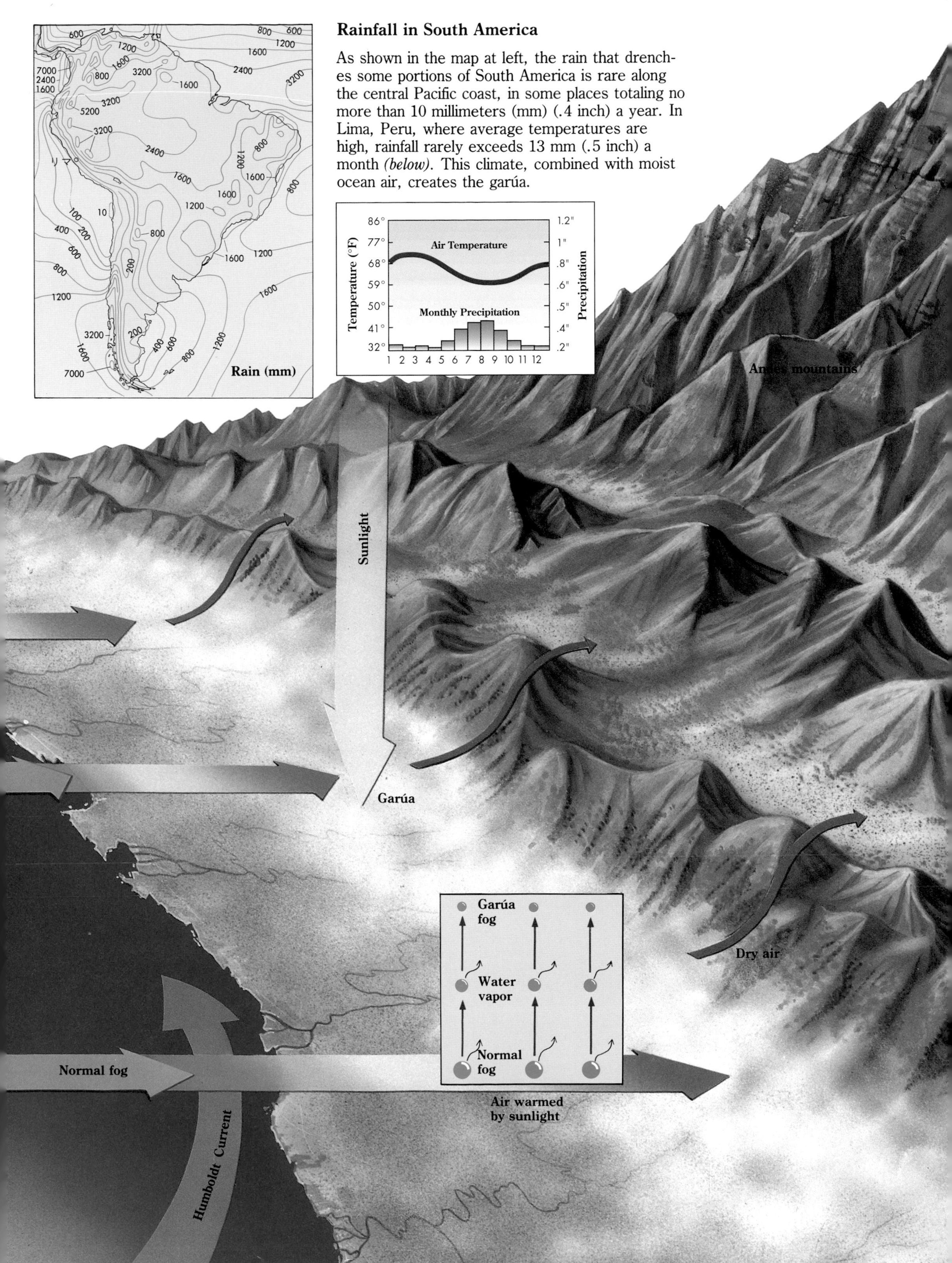

# What Causes Rain, Hail, and Snow?

The upper layers of high-floating cumulonimbus and altostratus clouds, where temperatures are well below freezing, are mostly made of ice crystals. Because the temperature is slightly higher in the middle layers, ice crystals rising and falling on air currents bump into supercooled water droplets. When these droplets attach themselves to the ice crystals, they form larger crystals heavy enough to fall through the cloud's ascending air currents.

On the way down, the crystals collide with other cloud particles and grow even larger. If the ground temperature is below freezing, the crystals fall as snow. If the air near the ground is warm, they become raindrops. When the ascending air currents inside the cloud are very strong, the ice crystals may rise and fall several times within the cloud. As they rise and fall, these ice crystals continue to grow and eventually become heavy enough to fall as hail. One of the largest pieces of hail ever reported fell in Coffeyville, Kansas, in 1970. It was nearly 6 inches wide and weighed 1⅔ pounds.

-4° F.

### The path of precipitation

This model of a cumulonimbus cloud formation *(right)* shows the path of air currents as they carry warm, vapor-laden air up to cooler altitudes and return with rain, snow, or hail.

Snow

32° F.

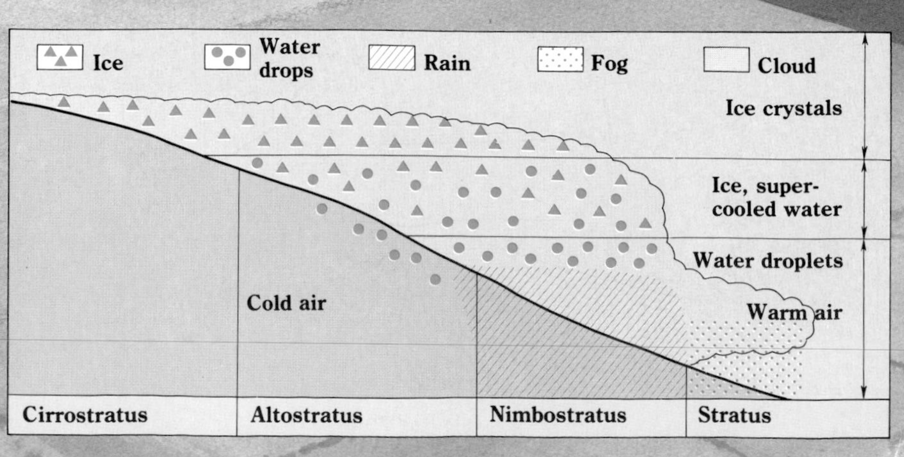


### Rain, snow, or hail

The cloud layers with the lowest temperatures *(graph, left)* will contain particles mostly in the form of ice crystals. At the slightly higher temperatures in lower layers, the ice mixes with water droplets to form crystals large enough to fall as rain, snow, or, if the conditions are right, hail.

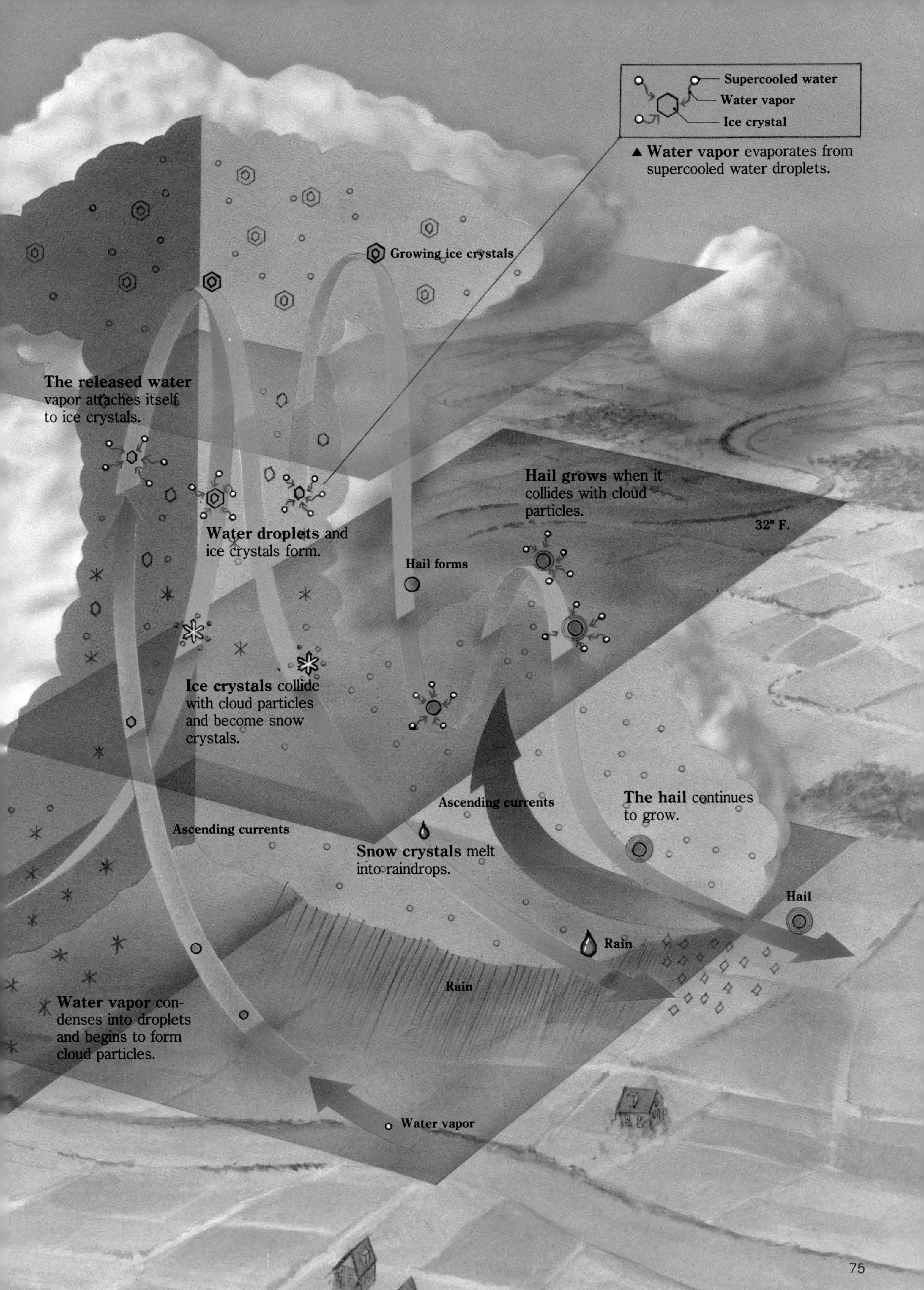
Supercooled water
Water vapor
Ice crystal
▲ **Water vapor** evaporates from supercooled water droplets.
Growing ice crystals
**The released water** vapor attaches itself to ice crystals.
**Hail grows** when it collides with cloud particles.
32° F.
**Water droplets** and ice crystals form.
Hail forms
**Ice crystals** collide with cloud particles and become snow crystals.
Ascending currents
Ascending currents
**The hail** continues to grow.
**Snow crystals** melt into raindrops.
Hail
Rain
Rain
**Water vapor** condenses into droplets and begins to form cloud particles.
Water vapor

# What Brings On a Tropical Squall?

The heavy afternoon rain shower that is a hallmark of the tropics is caused by high temperatures. In these hot, humid regions, the morning sun quickly warms and evaporates a large amount of moisture from the ocean surface and the land. The air near the ground is also warmed and rises steadily, carrying the evaporated water vapor with it. As the mass of warm air rises, it begins to cool. At a certain altitude the cooling water vapor condenses to form clouds. Because the strong sunlight is still heating the ground, the air currents containing water vapor will continue to rise and grow in strength. Soon, billowing cumulonimbus clouds develop. If these conditions persist, a short violent rainfall (squall) will disrupt the afternoon. On days with especially strong sunlight, the whole process can be repeated and produce a second afternoon squall.

**A banana leaf** protects a man from a squall on Indonesia's island of Bali.

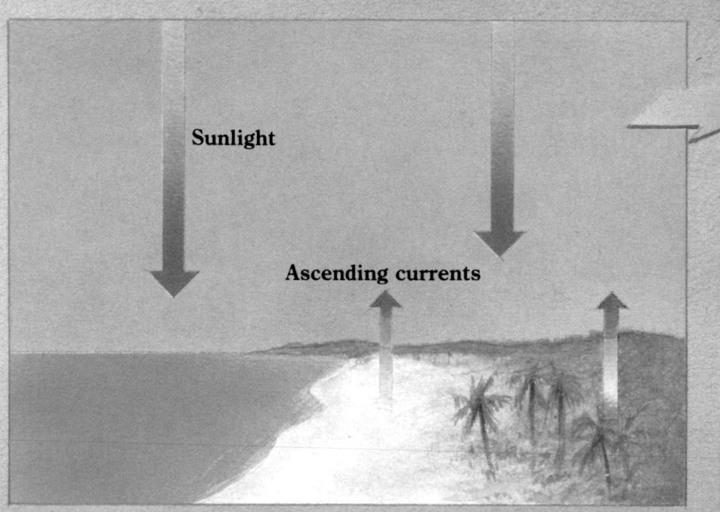


**Soon after sunrise,** strong tropical sunlight quickly warms the ground and the ocean surface, evaporating large amounts of water. The water vapor, caught up in a mass of warm air, is carried aloft on the ascending currents.

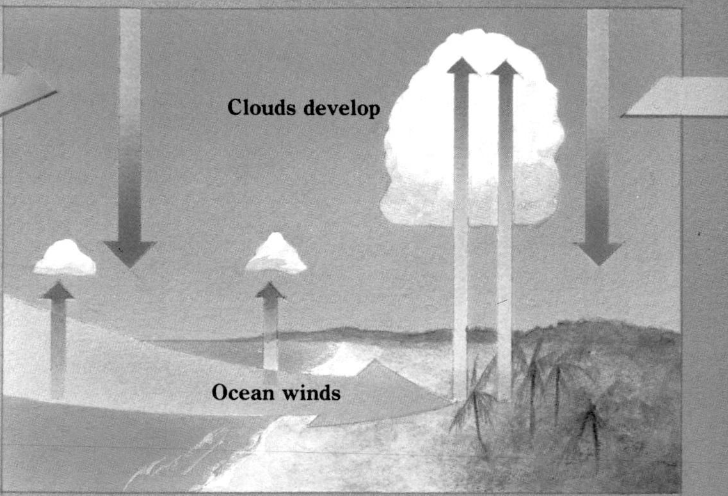


**At a certain height,** the rising air becomes saturated and forms clouds. The sunlight, now even stronger, produces increasingly violent ascending currents. The clouds turn into thunderclouds and are blown ashore by ocean winds.

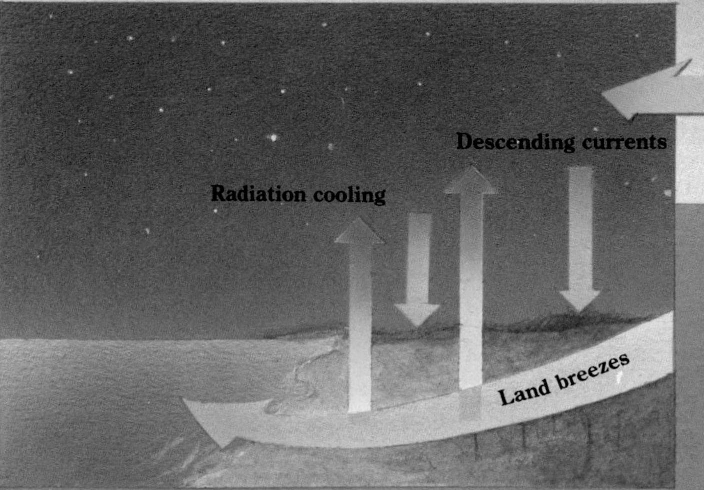


**During the cloudless night,** the ground radiates its heat and the temperature falls, producing cooler and more pleasant weather. A land breeze blows toward the ocean, causing descending currents over the land.

**Strong rains fall for only an hour** or two and, by evening, the clouds have disappeared. After the downpour, the temperature drops quickly. Humidity, affected by the heavy rains, greatly increases.

## Rainy days

This chart *(right)*, with the equator in the center, shows the number of rainy days during a one-year period around the world. Especially heavy rain falls in the Amazon Basin as well as in the South Pacific.

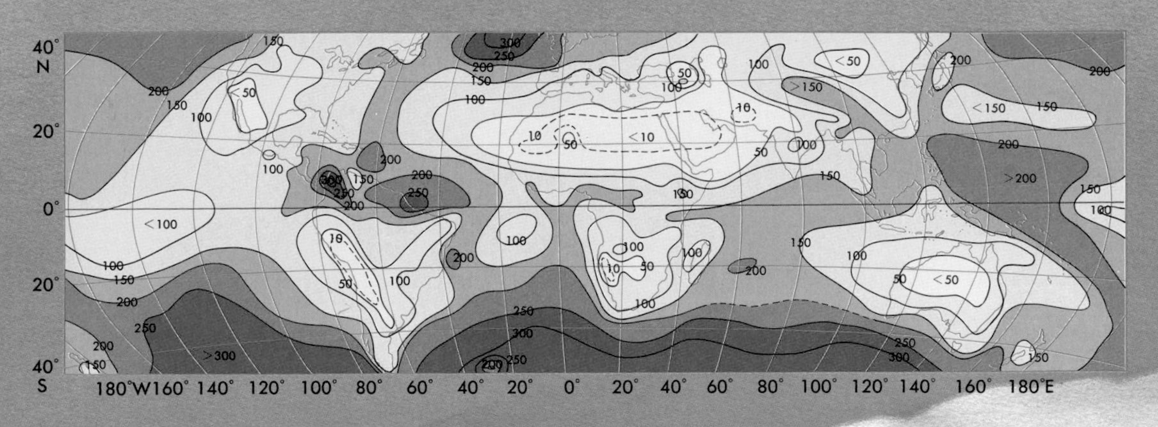


## Large drops of warm rain

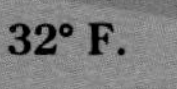


When the upper-atmosphere temperature remains above 32° F., the water droplets in the clouds *(right)* continue to grow instead of freezing into ice crystals. When they reach a certain size, the droplets become too heavy for the ascending currents and fall to the ground as the warm rain of a tropical squall.

# How Is Artificial Rain Made?

Researchers have come up with a handful of methods to encourage rain to fall from promising cumulonimbus clouds. In one technique, for example, airplanes drop particles of silver iodide or dry ice into the atmosphere, almost as if they were seeds. Water vapor attaches to these particles and, when the temperature is low enough, forms ice crystals. As these ice crystals fall toward the ground, they melt into raindrops.

Other methods involve sending particles up from smoke machines on the ground or sprinkling a foglike spray of water into the clouds from an airplane. No matter what the method used, none discovered so far can produce rain unless cloud conditions are already favorable. Artificial rainmaking methods have also been applied, with questionable success, to slow down and reduce the destructive power of hurricanes.

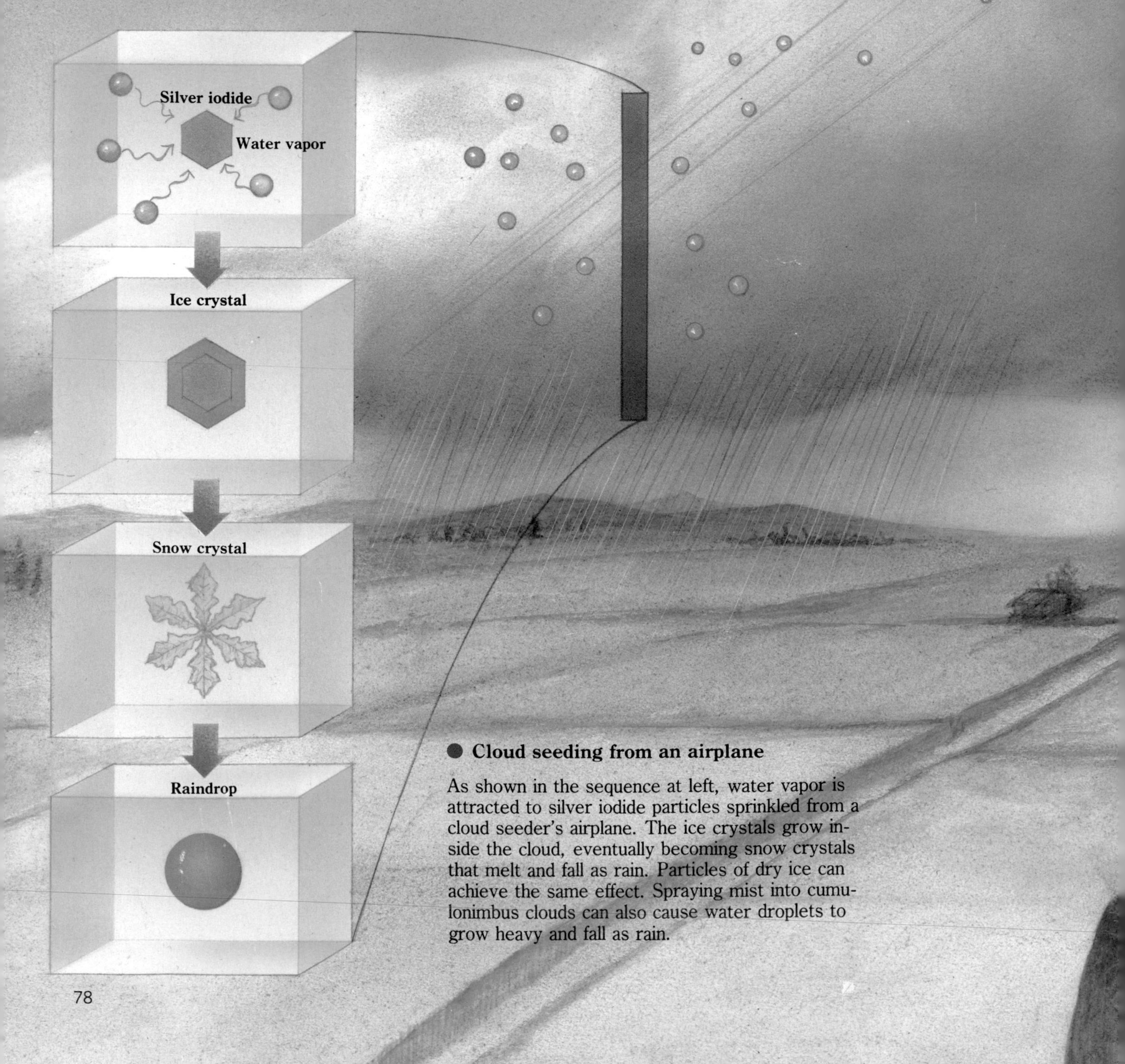

● **Cloud seeding from an airplane**

As shown in the sequence at left, water vapor is attracted to silver iodide particles sprinkled from a cloud seeder's airplane. The ice crystals grow inside the cloud, eventually becoming snow crystals that melt and fall as rain. Particles of dry ice can achieve the same effect. Spraying mist into cumulonimbus clouds can also cause water droplets to grow heavy and fall as rain.

## The best rainmaking conditions

During a 1962 study called the Sky Water Project, the U.S. government attempted to determine the most favorable conditions for producing artificial rain. Scientists working on the project focused on a dry belt extending from Texas to California. Among their findings was evidence that sprinkling silver iodide into clouds with temperatures between 14 and −9° F. could increase rainfall between 15 and 200 percent. As a result of the study, several privately owned rainmaking companies opened for business.

## Seeding from the ground

Smoke machines like this one *(left)* send silver iodide particles billowing into the atmosphere. This may increase rainfall by about 5 percent.

**Japan's Ogouchi Dam** *(above)*, built to store Tokyo's drinking water, is used during dry periods for experiments in artificial rain.

# What Happens in an Ice Storm?

When rain falls through a layer of subfreezing air near the ground, it may freeze and hit the ground as freezing rain or sleet. When the air temperature is right around freezing (32° F.) on a misty day, the weather may take an even more dramatic turn. Airborne water droplets, coming in contact with very cold objects such as tree branches or power lines, may freeze onto the structures. The ice may grow as more water freezes onto it in a process known as riming. On windy slopes, riming can transform trees into sculptures of ice and snow.

**Skiers move** among the strangely graceful "ice monster" formations that cloak the trees on Japan's Mt. Zao.

**Rime can form** beautiful, feathery shapes on branches.

**Tree**

**Cross section**

**Ice develops on the trees**

**Flow of the wind**

**Supercooled water droplets**

**Icing**

### Icing on mountain slopes

Wind-swept water droplets smashing into the branches of pine trees on a mountainside can form strange icy sculptures, as shown above. After the water droplets freeze into ice, they become covered with snow. Winds sweeping around the structures curl around to deposit further layers on the backs of the trees.

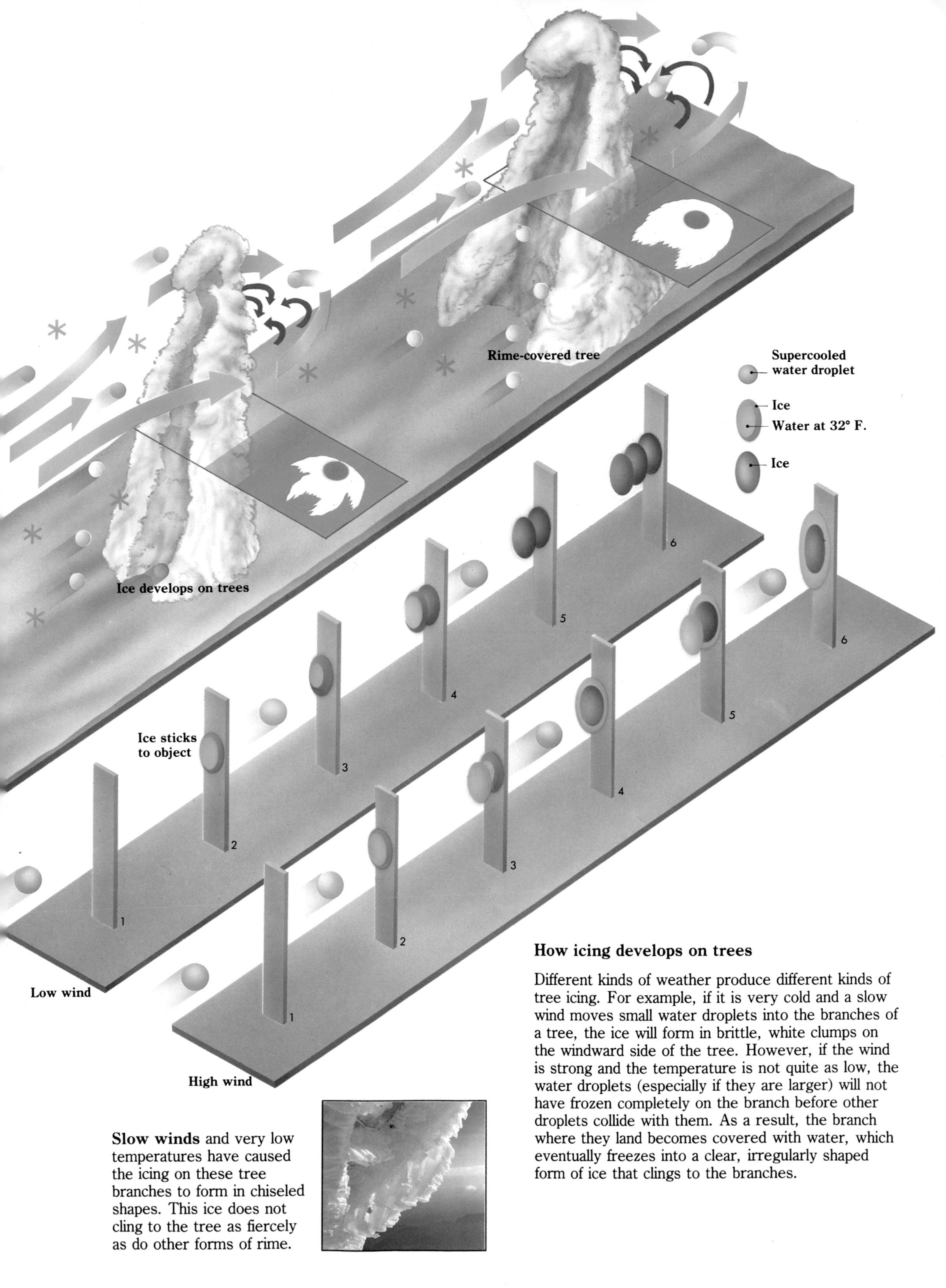


## How icing develops on trees

Different kinds of weather produce different kinds of tree icing. For example, if it is very cold and a slow wind moves small water droplets into the branches of a tree, the ice will form in brittle, white clumps on the windward side of the tree. However, if the wind is strong and the temperature is not quite as low, the water droplets (especially if they are larger) will not have frozen completely on the branch before other droplets collide with them. As a result, the branch where they land becomes covered with water, which eventually freezes into a clear, irregularly shaped form of ice that clings to the branches.

**Slow winds** and very low temperatures have caused the icing on these tree branches to form in chiseled shapes. This ice does not cling to the tree as fiercely as do other forms of rime.

# 4

# Atmospheric Pressure

Although the atmosphere is invisible and seemingly weightless, in fact it has heft and substance. At the surface of the Earth, the overlying air mass exerts a pressure of about 15 pounds per square inch. Most meteorologists now measure pressure in units called millibars; the standard pressure at sea level is 1013 mb.

But atmospheric pressure is not uniform, and the differences between pressures in different locations help drive the weather machine. In the Northern Hemisphere, for instance, winds blow outward from the center of high-pressure regions in a clockwise direction. High-pressure regions are generally associated with fair weather. In low-pressure areas in the Northern Hemisphere, the winds are usually warmer and blow in a counterclockwise direction. Local regions of low pressure may collide with a high-pressure system to produce a weather front, where thunderstorms are born. Low-pressure centers also form over tropical oceans. Fed by the energy produced by the evaporation and condensation of warm ocean waters, these tropical depressions may grow into huge storm systems known as hurricanes in the Atlantic Ocean and typhoons in the Pacific.

The swirling winds of a hurricane show up clearly in a satellite photograph. Feeding on the energy of warm tropical waters, a low-pressure depression may grow into a violent storm system hundreds of miles in diameter.

# What Is a High-Pressure System?

No one can understand the strength of winds or the birth of a storm without following high- and low-pressure systems. On a weather map, lines connecting areas of equal atmospheric pressure are called isobars. A typical map might connect all areas in which the pressure measures 1000 millibars. Some isobars form closed circles around high-pressure regions, indicating the presence of a high-pressure system.

Winds blow outward from the center of a high-pressure system, cutting across the isobars in a clockwise direction in the Northern Hemisphere and counterclockwise in the Southern Hemisphere. These winds blow downward from a high altitude, carrying away any clouds in the region; as a result, high-pressure systems usually bring fair weather. High-pressure systems may consist of warm or cold air masses. They often linger in one spot, although some, moving from west to east, are known as migratory high-pressure systems.

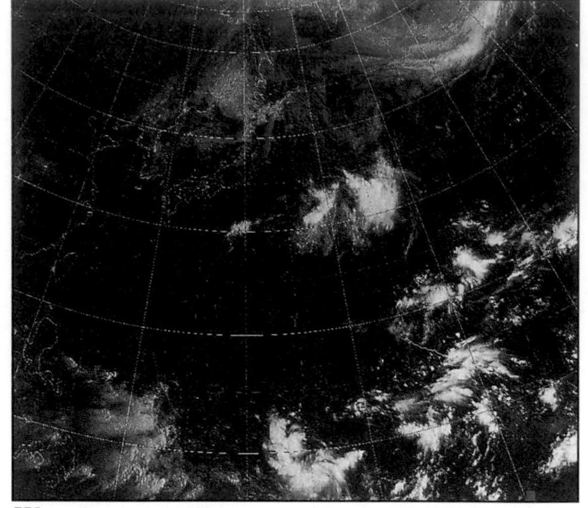

**Weather satellites** provide detailed information on the movement of weather systems such as these over the western Pacific Ocean.

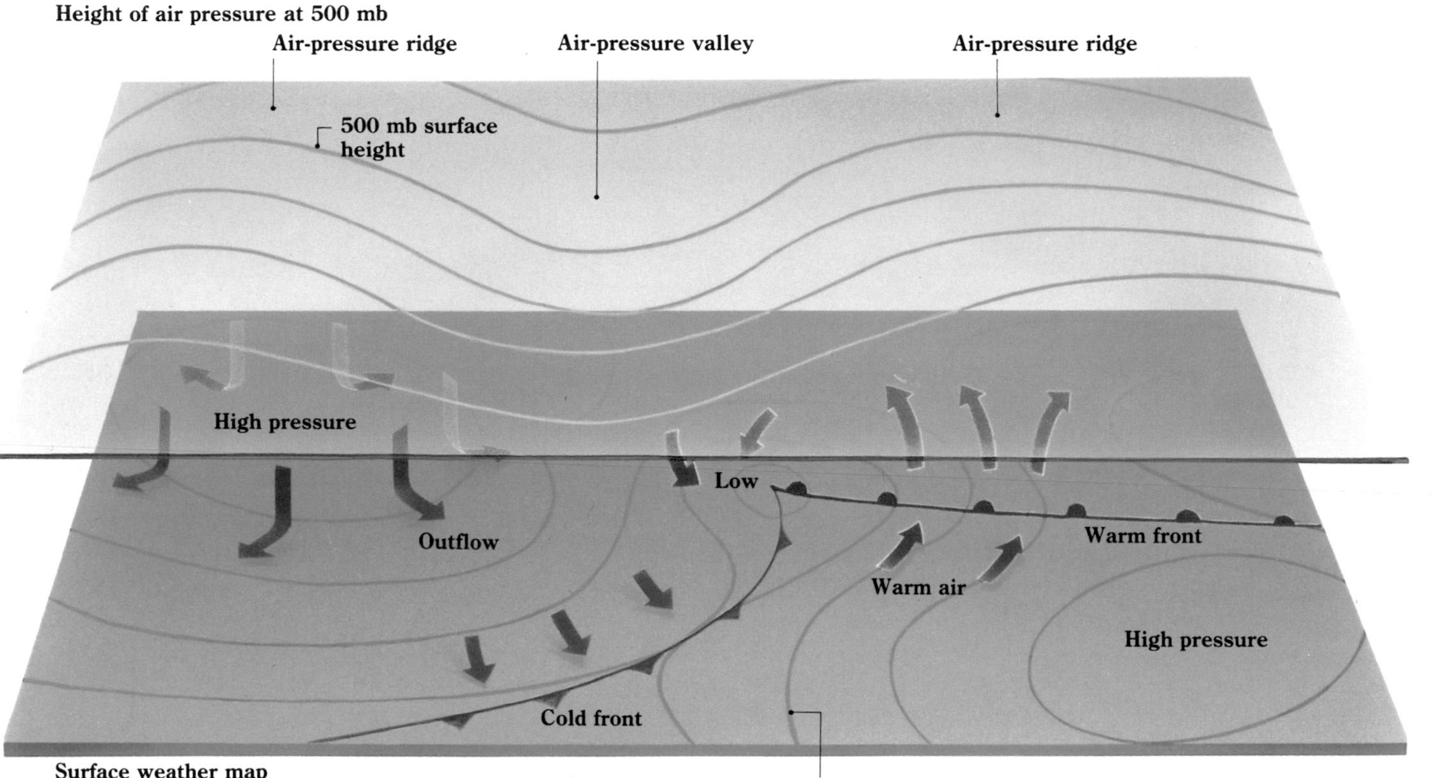


## Tracking a front

As high-altitude westerlies blow from west to east, a high-pressure "ridge" may develop along the leading edge of the system. To the east of the ridge lies a low-pressure "valley." Air blows away from the high-pressure system and into the low-pressure valley. Because there is now less air in the upper atmosphere, air converging on the low-pressure area near the surface is swept up into ascending air currents to the east of the ridge. These ascending currents often form cumulonimbus clouds, creating thunderstorms along the leading edge of the front. The horizontal motion of high-pressure systems is thus accompanied by rapid vertical movements of air currents. This dance of air in the upper sky affects local weather conditions.

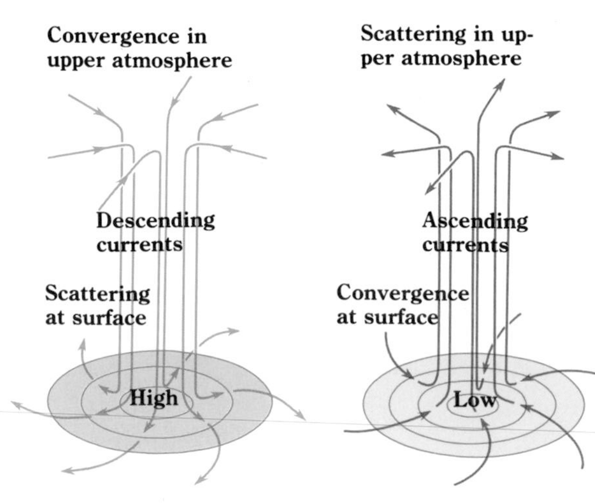

## Two types of high pressure

Migratory high-pressure systems move in a generally predictable pattern, controlled by the high-altitude jet streams. Stationary high-pressure systems, which may include either warm or cold air, tend to remain in one place.

### Warm high pressure

Near the equator, air warmed by the sun rises and spreads to the north and south. At midlatitudes, it descends to form high-pressure systems. The high column of air associated with these warm high-pressure systems extends to the tropopause, the upper boundary of the troposphere. The Earth is encircled by subtropical high-pressure belts formed in this manner. The Pacific high, shown at right, that spreads out near Japan each summer is an example of this process.

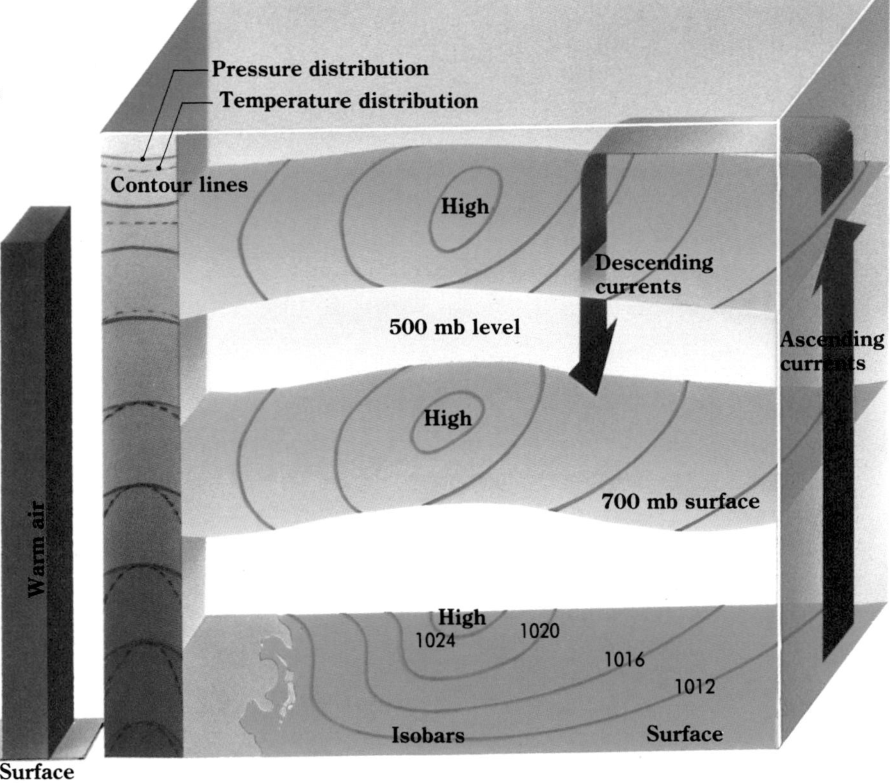


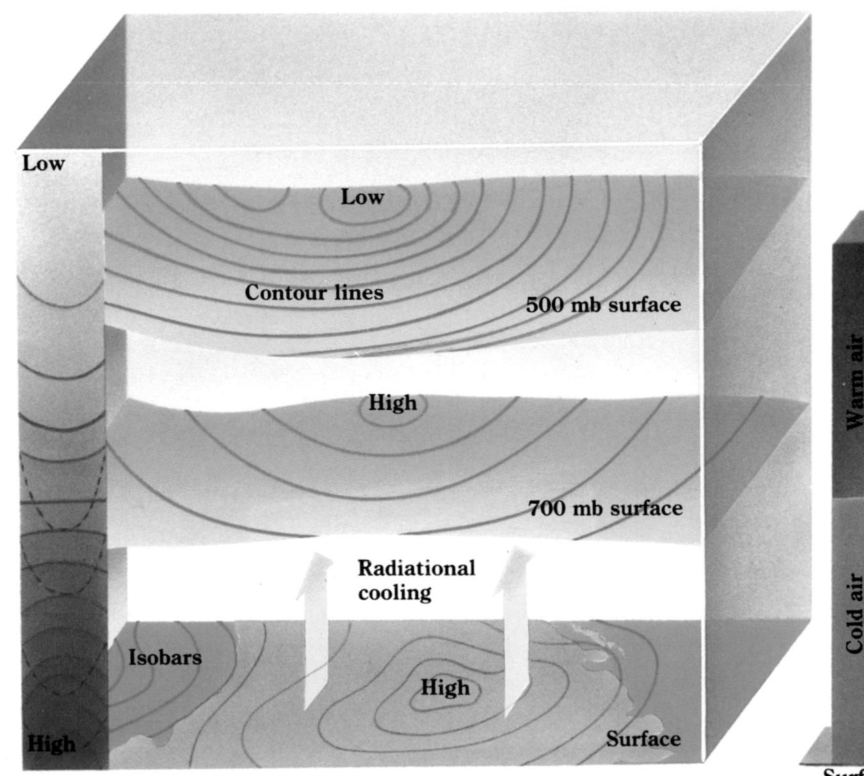


### Cold high pressure

During the winter, air at high latitudes receives little sunlight and becomes cold. Cold air is heavier than warm air and descends toward the surface, creating a cold high-pressure system. Cold high-pressure systems form over continental landmasses, where there is no warm ocean air. Such systems typically form each winter over Siberia and Canada.

### Migratory high pressure

Migratory high-pressure systems move in a relatively definite direction over short cycles. In Japan, for example, migratory high-pressure systems moving eastward from the Asian mainland during spring and autumn create cycles of fair and cloudy weather, each lasting several days.

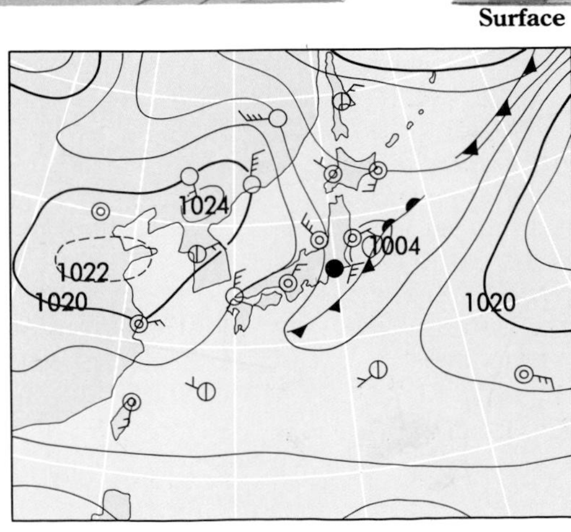


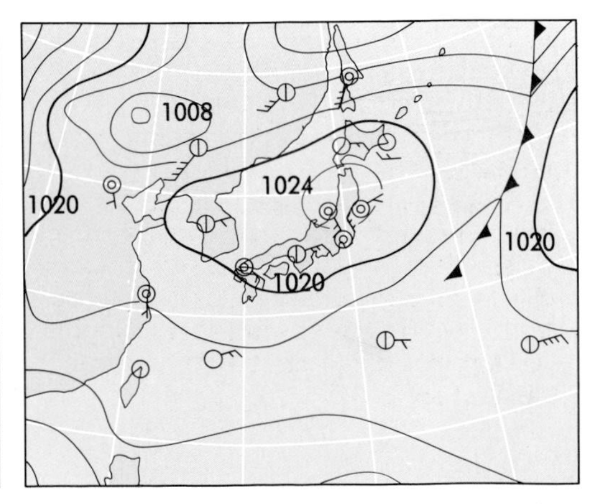

# What Is a Low-Pressure System?

A low-pressure system is a region of low atmospheric pressure marked by rising air currents and, frequently, rain. Most of these systems form as waves on the leading edge of cold fronts moving west to east. To replace the air that is swept upward, winds blow into a low-pressure system from outside, moving in a counterclockwise direction in the Northern Hemisphere.

Warm low-pressure systems develop in mid- and high latitudes, while tropical low-pressure systems, which may give birth to typhoons and hurricanes, form over the ocean in regions near the equator. Low-pressure systems may also develop because of localized warming inland, while topographical lows often form on the leeward side of mountains.

**Clouds swirl in a counterclockwise eddy** around the center of this low-pressure system in the western Pacific.

**A low-pressure system evolves**

**1 Generation period.** Westerly winds in the upper atmosphere create a wave known as a low-pressure valley or trough. To the east of the trough, warm air climbs above the cold air to create a warm front. To the west, cold air descends in a cold front. A whirlpool-shaped current (known as a cyclone) forms around the wave, generating a low-pressure system. (The white dotted line shows where the cold front is located relative to the trough.)

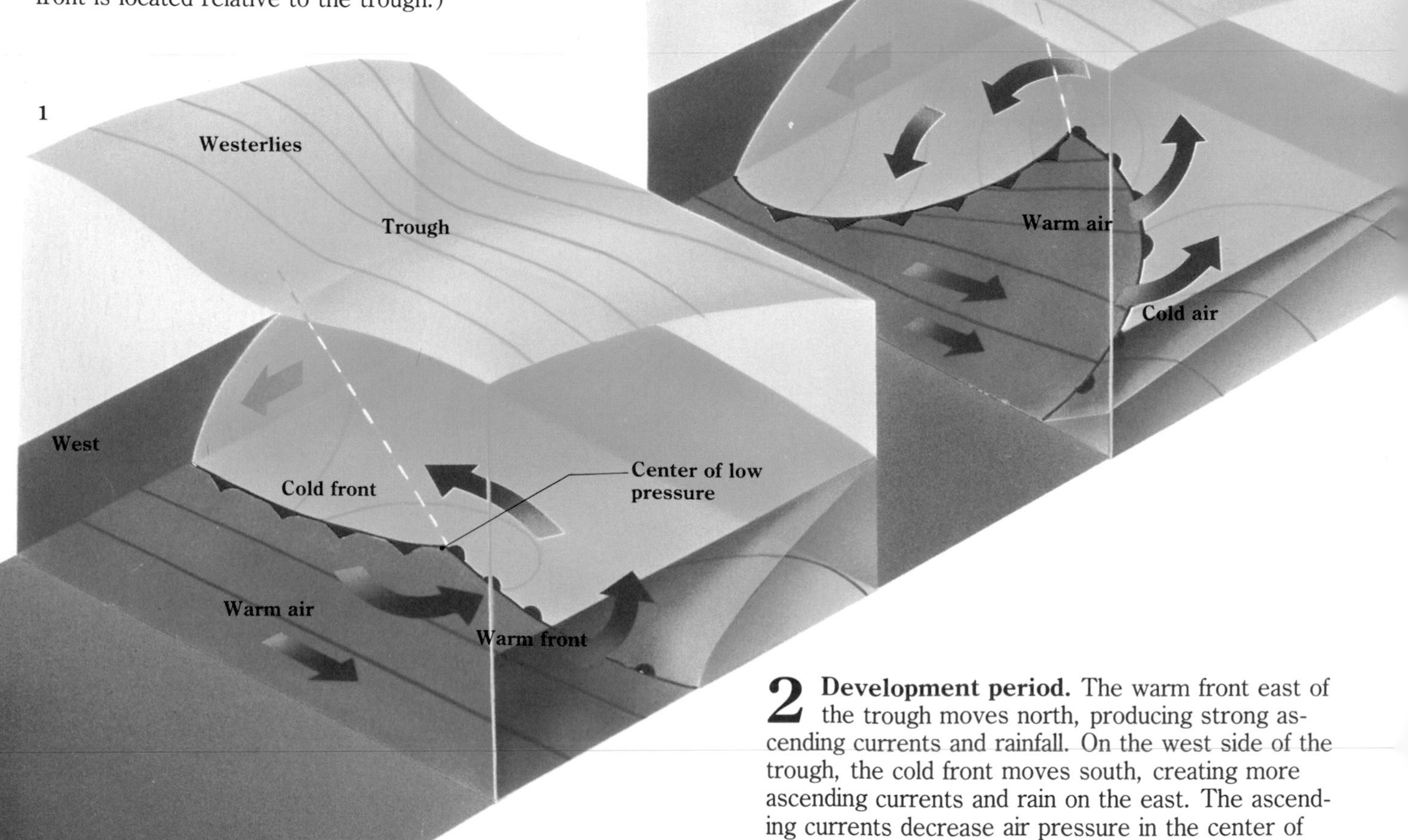


**2 Development period.** The warm front east of the trough moves north, producing strong ascending currents and rainfall. On the west side of the trough, the cold front moves south, creating more ascending currents and rain on the east. The ascending currents decrease air pressure in the center of the system, and the cyclone grows stronger.

**3** **Peak period.** The low-pressure system reaches its peak of intensity as the low-pressure trough meets the cold front. A warm front moves northwest from the center of the system while a cold front moves southwest. Because the cold front moves more quickly, it catches up with the edge of the warm front and pushes the warm air to higher altitudes. Cold air blankets the ground as an occluded front begins to form.

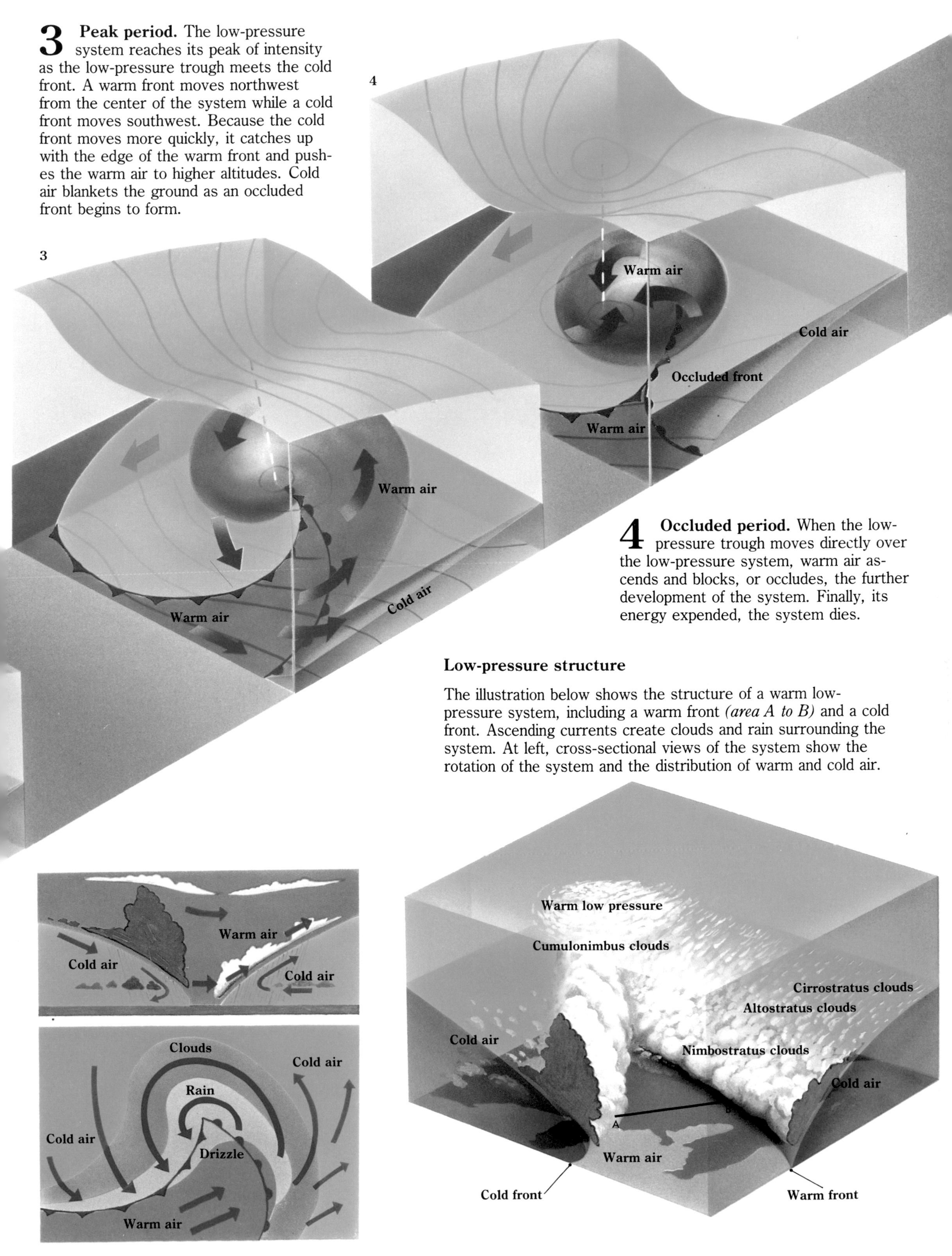


**4** **Occluded period.** When the low-pressure trough moves directly over the low-pressure system, warm air ascends and blocks, or occludes, the further development of the system. Finally, its energy expended, the system dies.

### Low-pressure structure

The illustration below shows the structure of a warm low-pressure system, including a warm front *(area A to B)* and a cold front. Ascending currents create clouds and rain surrounding the system. At left, cross-sectional views of the system show the rotation of the system and the distribution of warm and cold air.

# What Causes Heavy Downpours?

Most areas of the world that receive rain are familiar with the occasional torrential downpour, often a result of a thunderstorm. Some regions, especially the Pacific coast of Asia, experience heavy rains every summer.

Rains such as these occur when large amounts of water vapor are carried into an area and caught up in strong rising air currents, so that the vapor condenses. Over the United States during the summer months, the low-level jet stream usually dips southward and imports warm, moist air from the Gulf of Mexico in a shape called a "wet tongue." When this moist air collides with a cold front, a low-pressure system may form. The ascending currents of the cold front sweep the moist air upward, forming cumulonimbus clouds and bringing heavy rain.

**Summer storms**

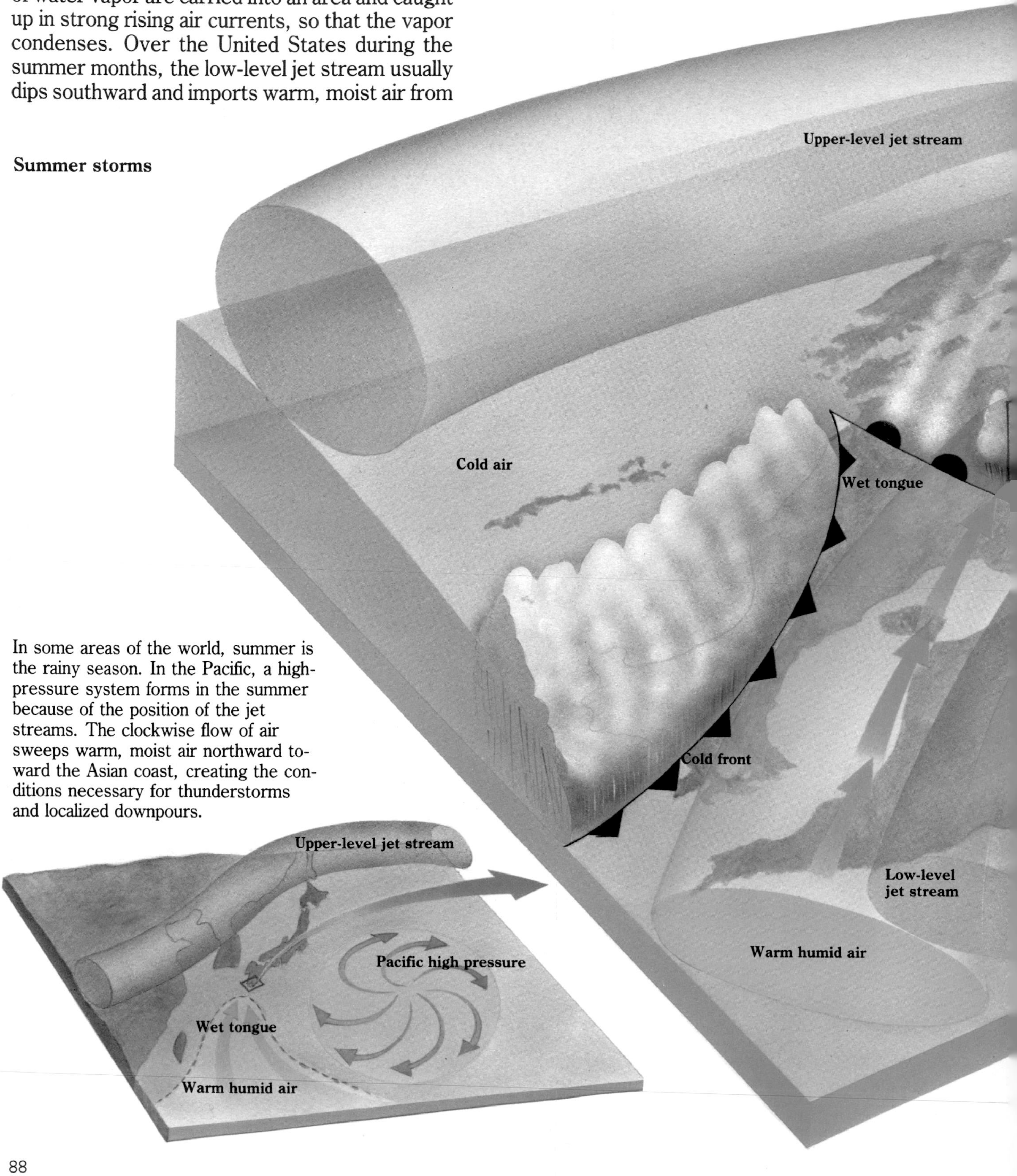

In some areas of the world, summer is the rainy season. In the Pacific, a high-pressure system forms in the summer because of the position of the jet streams. The clockwise flow of air sweeps warm, moist air northward toward the Asian coast, creating the conditions necessary for thunderstorms and localized downpours.

### The rainy season

During the Asian summer, the continent's landmass is heated so intensely that the rising air creates a large area of low pressure. This brings in cool, moisture-laden wind from the ocean. Rising rapidly on the heated currents from the mainland, the water vapor condenses into heavy rains throughout the season.

**Villagers on Borneo** in Southeast Asia endure the island's soaking seasonal rains.

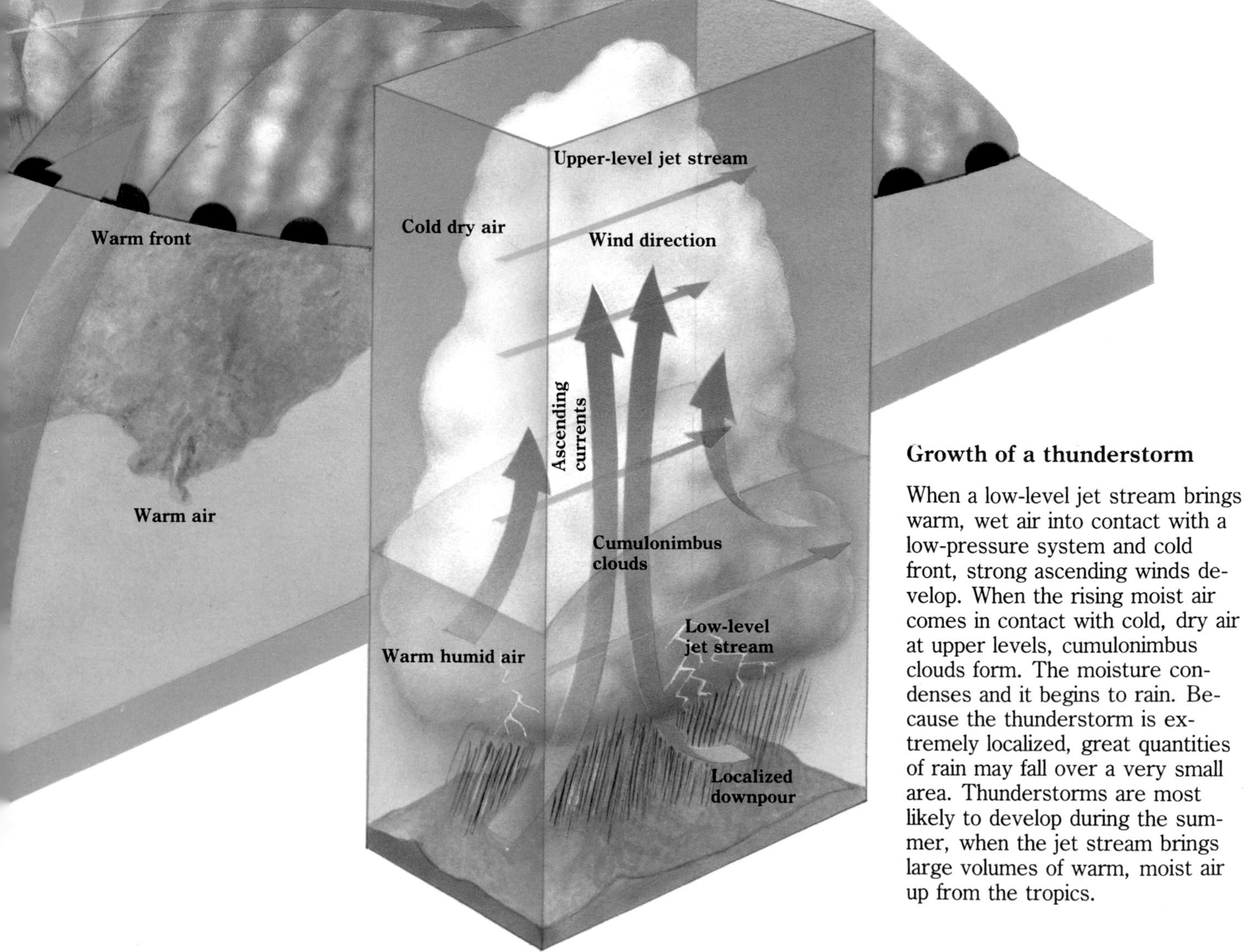


### Growth of a thunderstorm

When a low-level jet stream brings warm, wet air into contact with a low-pressure system and cold front, strong ascending winds develop. When the rising moist air comes in contact with cold, dry air at upper levels, cumulonimbus clouds form. The moisture condenses and it begins to rain. Because the thunderstorm is extremely localized, great quantities of rain may fall over a very small area. Thunderstorms are most likely to develop during the summer, when the jet stream brings large volumes of warm, moist air up from the tropics.

# Why Do Hurricanes Happen?

Hurricanes and their Pacific counterparts, typhoons, are born over the warm waters of tropical oceans. As the summer sun heats the water, the warm water vapor rises into the atmosphere, forming cumulonimbus clouds. Rich with moisture and energy from the ocean, these clouds may combine to form vast, low-pressure whirlpools. Strong winds start the clouds spinning; in the Northern Hemisphere, the spin is counterclockwise because of the Coriolis effect. Such tropical depressions generally move west then northwest, gathering energy and moisture as they go.

▲ **A towering cumulonimbus whirlpool** forms above the tropical ocean. This signals the birth of a hurricane.

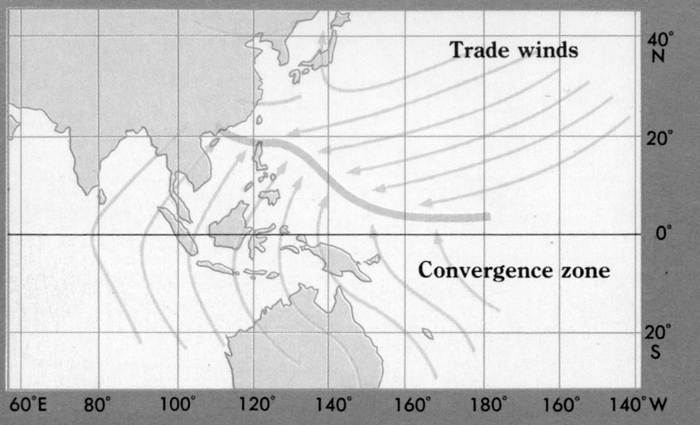


▲ **Colliding air currents** in the northern equatorial convergence belt give birth to typhoons in the western Pacific Ocean.

Northeast trade winds

Cumulonimbus clouds

Southeast trade winds

Ascending currents

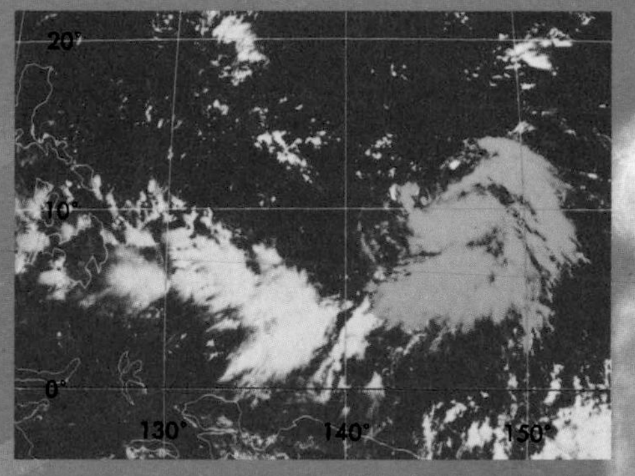


▲ **Low-pressure system *(yellow)***

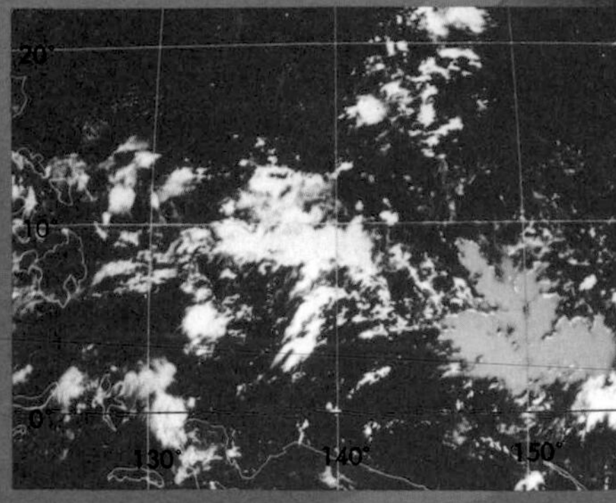


▲ **Cumulonimbus clouds *(yellow)***

**1 Strong ascending** currents develop as trade winds collide, forming groups of cumulonimbus clouds.

## Hurricane zones

Hurricanes and typhoons develop in zones from 5° to 20° latitude north and south of the equator, where the tropical ocean is about 80° F. or higher *(pink areas)*. About three times more of these storms form in the Northern Hemisphere than in the Southern. The storms generally move west, then north.

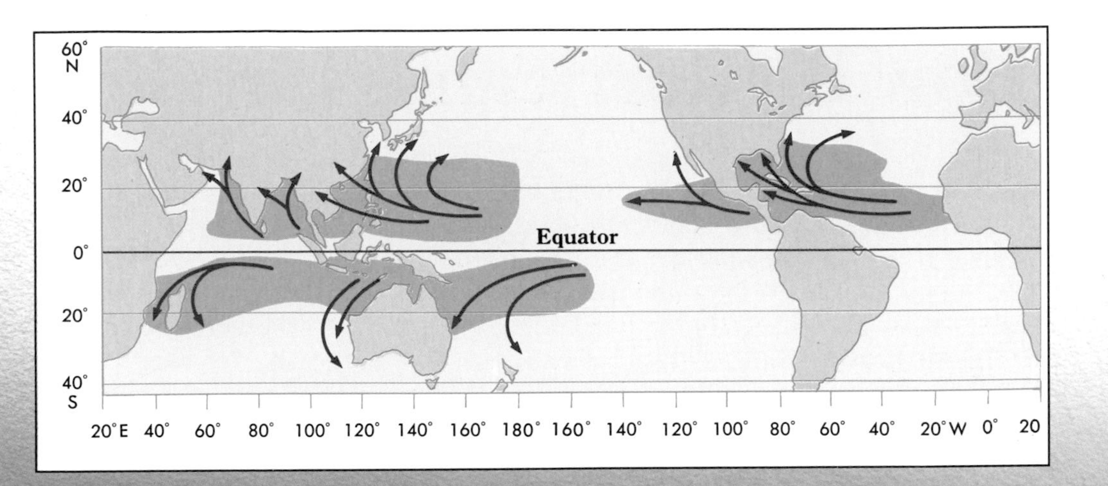


▲Hurricane

**3 As the storm** feeds on the energy and moisture of the tropical oceans, winds strengthen and a clearly defined "eye" forms in the middle of the whirlpool.

Hurricane

Outflow

Eye

Descending currents

Ascending currents

Tropical low pressure

Ascending currents

**2 As ascending currents** grow in the cumulonimbus formations, warm, moist air is drawn in. The low-pressure system at the core of the storm begins to rotate counterclockwise. More air is drawn in and the storm system expands. At the top of the storm, high-altitude winds pull the clouds clockwise.

### The eye of the storm

As wind blows toward the center of a storm, it is acted upon by competing forces. Centrifugal force pulls away from the center, while a "tilting force" caused by air pressure pushes inward. At the center of the storm, these forces are in balance, preventing more wind from blowing toward the center. A circular, low-pressure eye develops, creating a patch of clear sky in the center of the vast storm system.

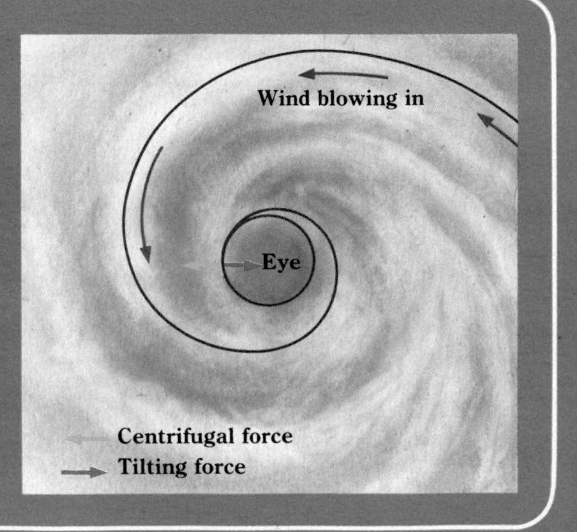

# How Do Hurricanes Develop?

When a tropical depression develops to the point at which the maximum sustained winds reach a velocity of 75 miles an hour, it is classified as a hurricane. In this stage, the storm is moving west or northwest, gathering additional energy from the warm tropical ocean. Evaporation of surface waters feeds water vapor into the ascending currents of the storm. Pressure continues to drop and the eye forms.

As the hurricane (or typhoon) reaches its peak intensity, air currents formed by the Bermuda or Pacific high-pressure areas shift its course northward. Here, the ocean waters are cooler. There is less evaporation, less water vapor, and less energy to feed the storm. If the storm hits land, the supply of water vapor is cut off entirely. As the hurricane or typhoon continues to move north, its winds begin to diminish and the eye disintegrates. Topographical features such as mountains may also contribute to the breakup of the storm. Finally, its energy spent, the storm becomes a harmless warm low-pressure system.

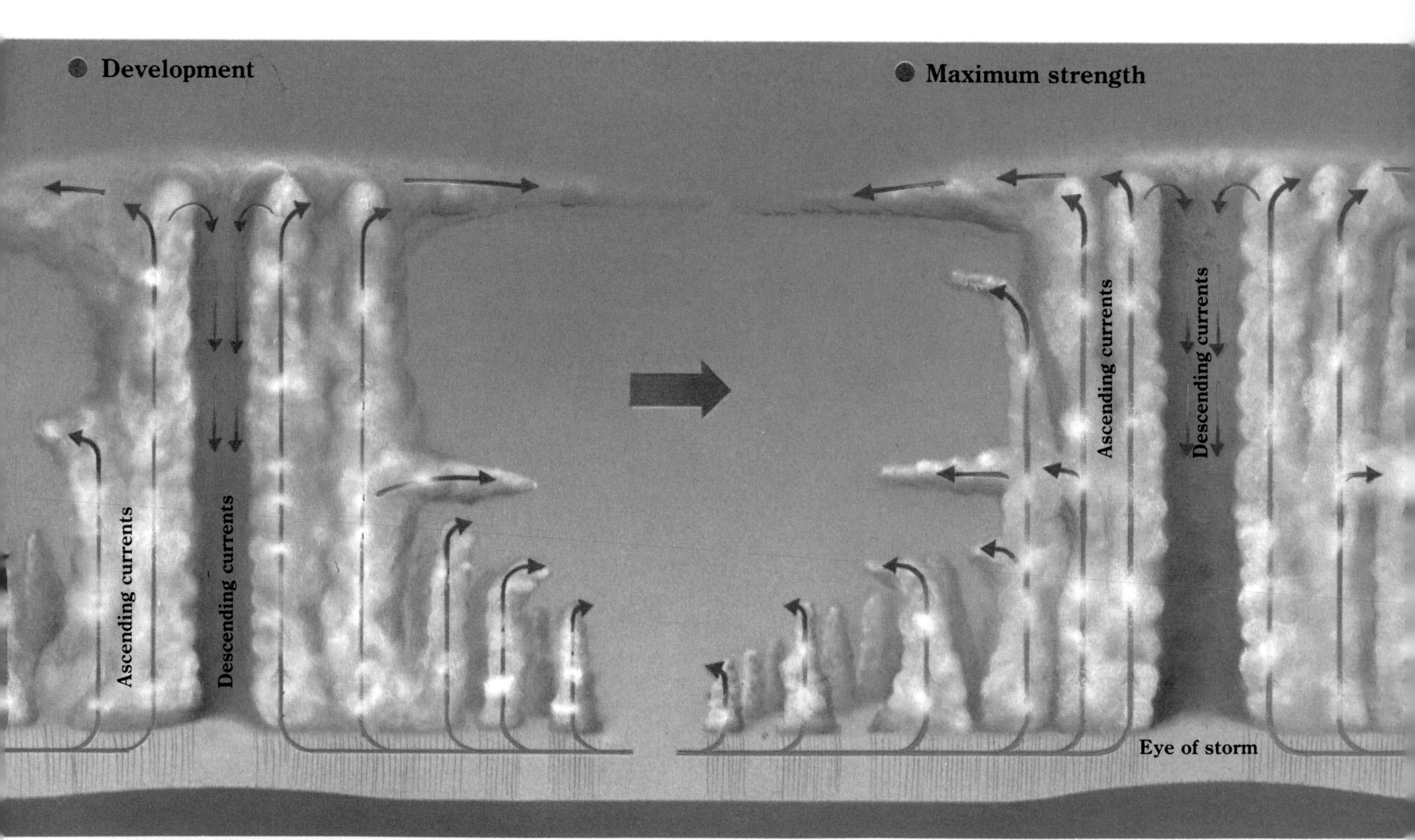


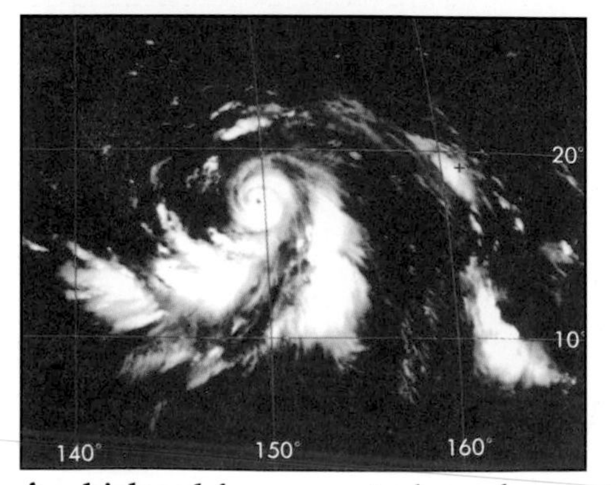


**A whirlpool** forms as a typhoon develops over the tropical ocean.

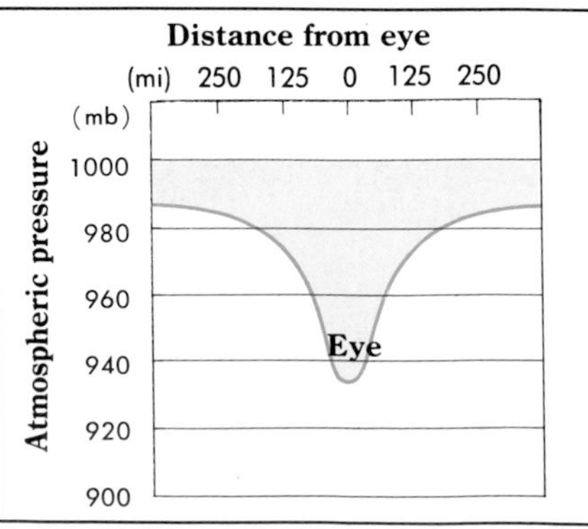


**Atmospheric pressure** drops at the center of the storm.

**The storm coils tightly,** forming an eye as it reaches maximum strength.

## The life of a typhoon

A typical Pacific typhoon forms in the warm tropical waters east of the Philippines. It may move westward and strike the Chinese mainland, or veer to the north and approach Japan. The storm's path is determined as it moves around the western edge of the Pacific high-pressure system. The maps at right show the tracks of typhoons. At far right, isobars show the waxing and waning of the typhoon's strength. The storm is at its peak strength when there is a maximum number of isobars with a minimum distance between them. As the storm begins to disintegrate, pressure rises and the isobars become fewer and more widely spaced.

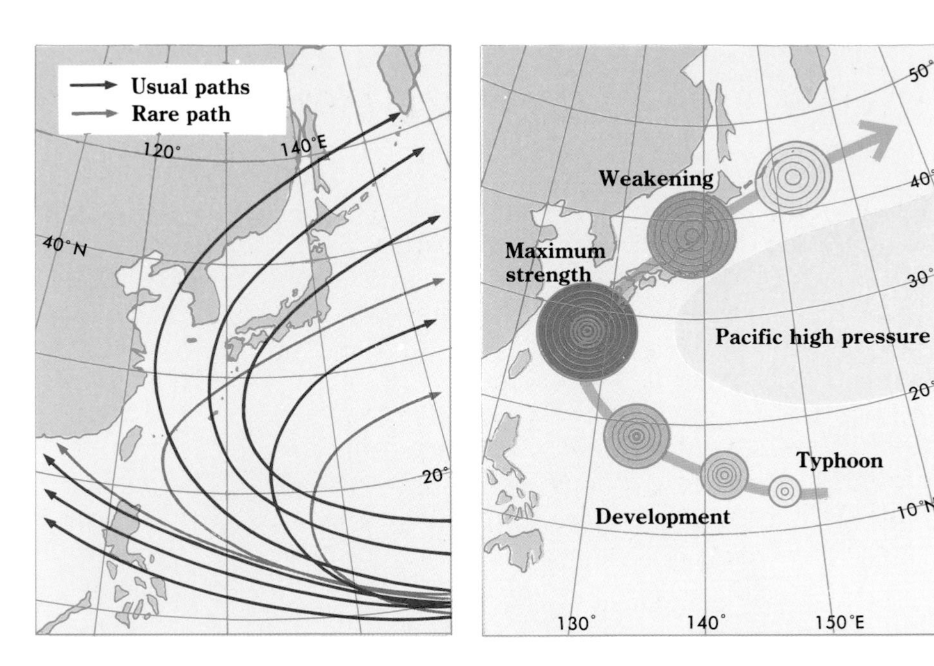


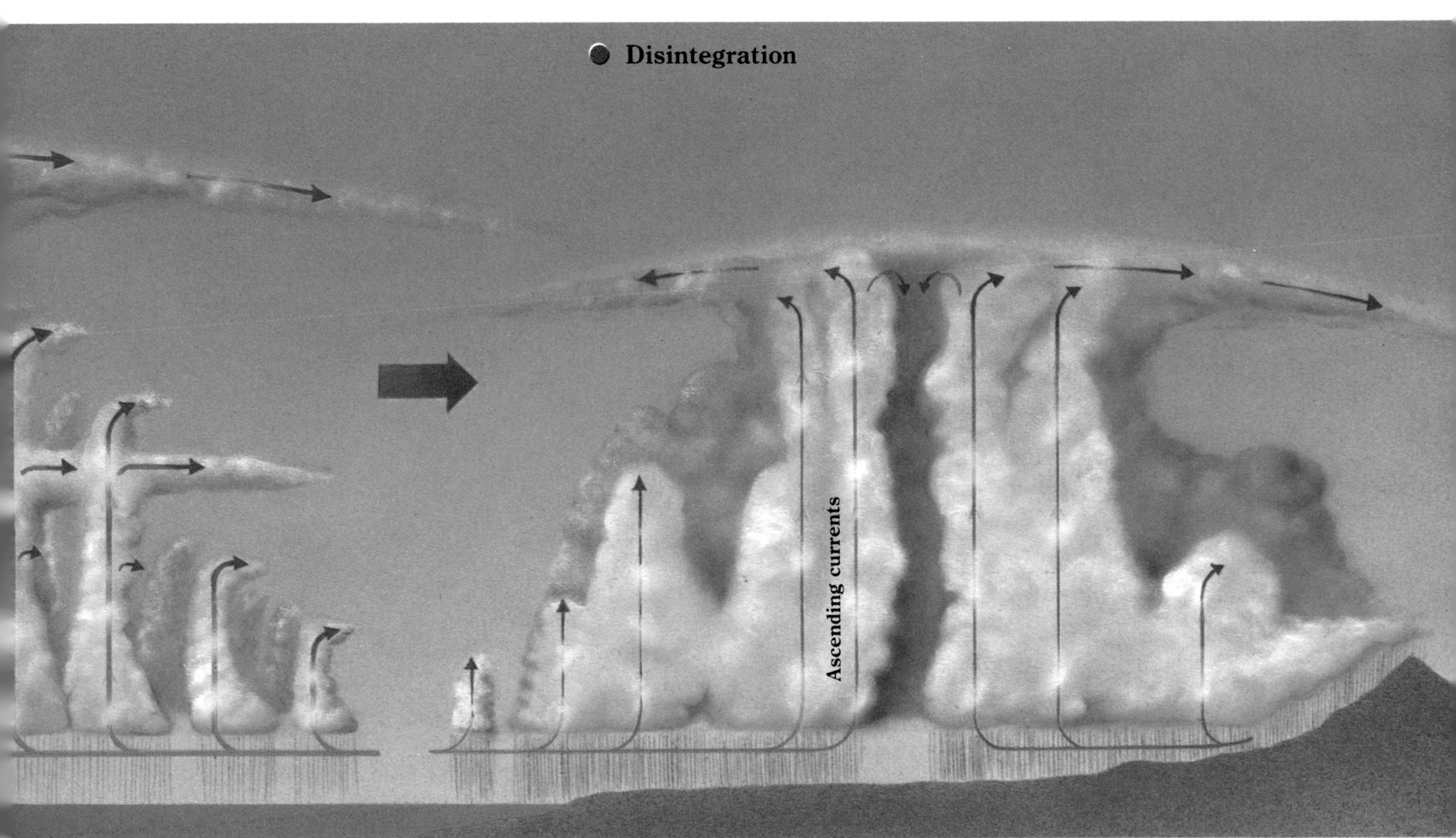


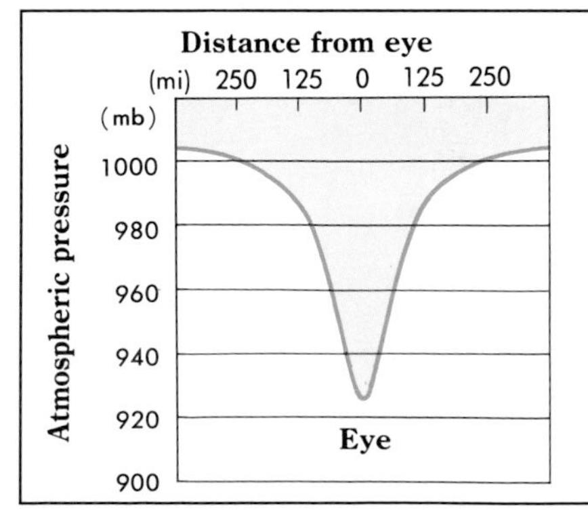


**Pressure plummets** to a minimum in the eye of the storm.

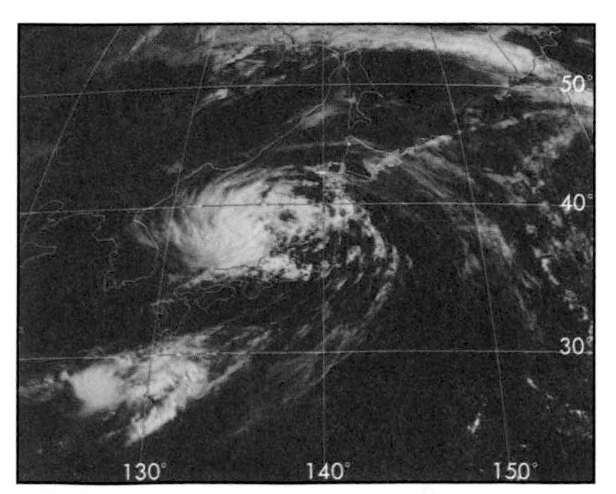


**After striking land,** the storm rapidly breaks up.

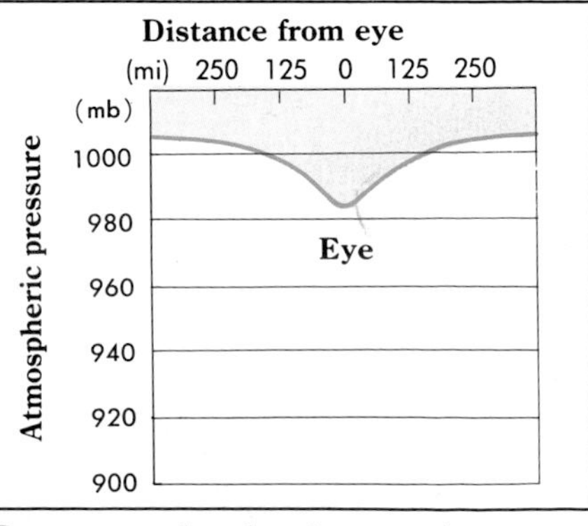


**Pressure slowly rises** at the center as the storm dies.

# How Are Hurricanes Observed?

Before the development of modern methods for tracking weather, people had little warning that a dangerous storm was bearing down upon them. Today, storms are watched carefully from the moment they first begin to form in the tropical ocean. Weather satellites in geostationary orbit above the equator can keep a constant watch on the areas where storms are born. When a storm is spotted, "hurricane hunter" weather observation planes are dispatched to the area. The planes fly directly into the growing storm, at significant risk to the pilot and scientists aboard, to collect data on pressure, temperature, wind speed, wind direction, and rainfall. As the storm approaches land, radar also tracks it continuously. Each change in the storm's path is noted in order to predict the storm's exact landfall and give area residents the earliest possible warning.

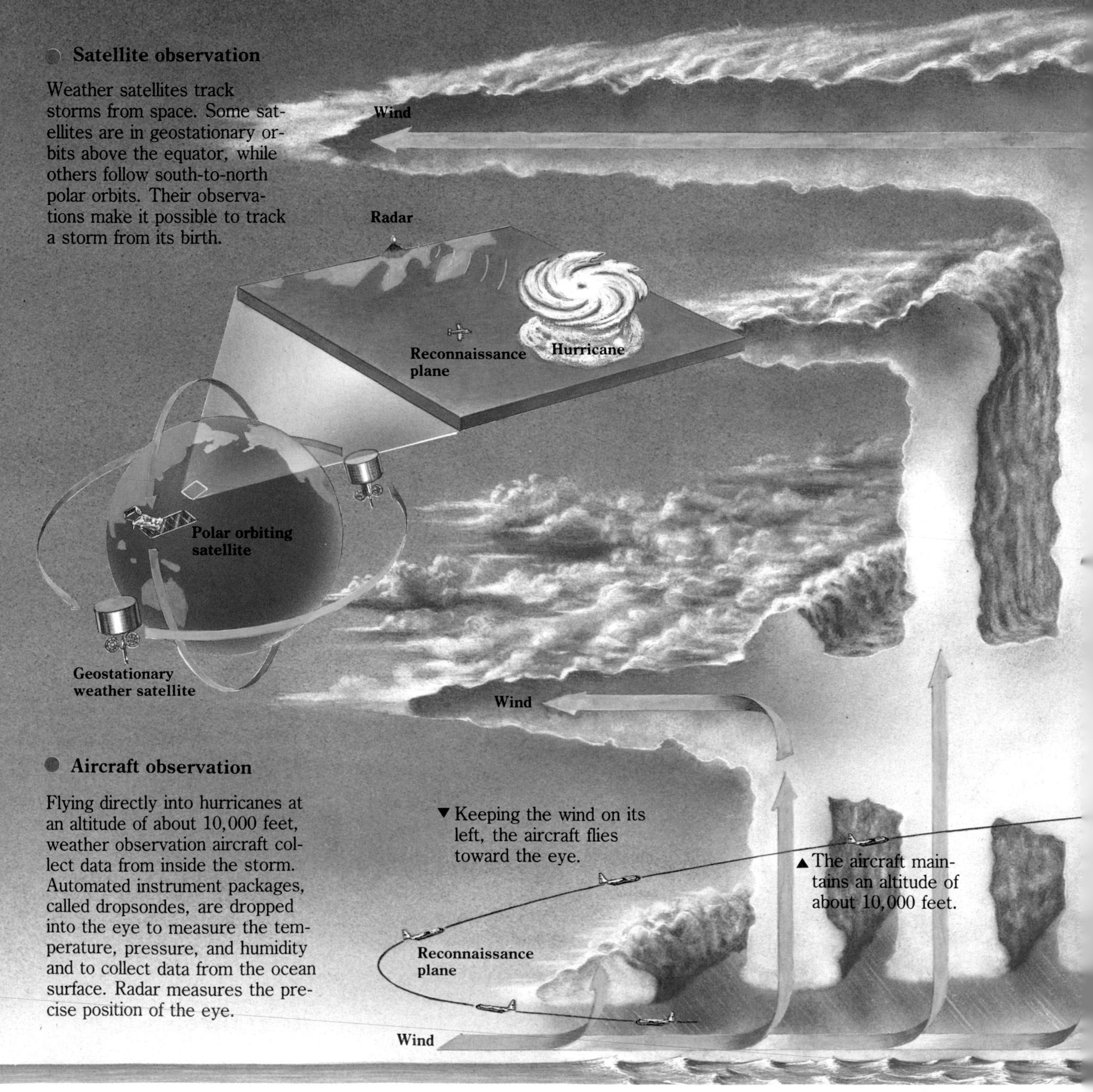


**Satellite observation**

Weather satellites track storms from space. Some satellites are in geostationary orbits above the equator, while others follow south-to-north polar orbits. Their observations make it possible to track a storm from its birth.

**Aircraft observation**

Flying directly into hurricanes at an altitude of about 10,000 feet, weather observation aircraft collect data from inside the storm. Automated instrument packages, called dropsondes, are dropped into the eye to measure the temperature, pressure, and humidity and to collect data from the ocean surface. Radar measures the precise position of the eye.

▲ **Flying into the heart of the storm,** pilots and scientists risk their lives to gather vital data.

▲ **Towering walls of clouds** encircle the sunlit calm in the peaceful eye of the storm.

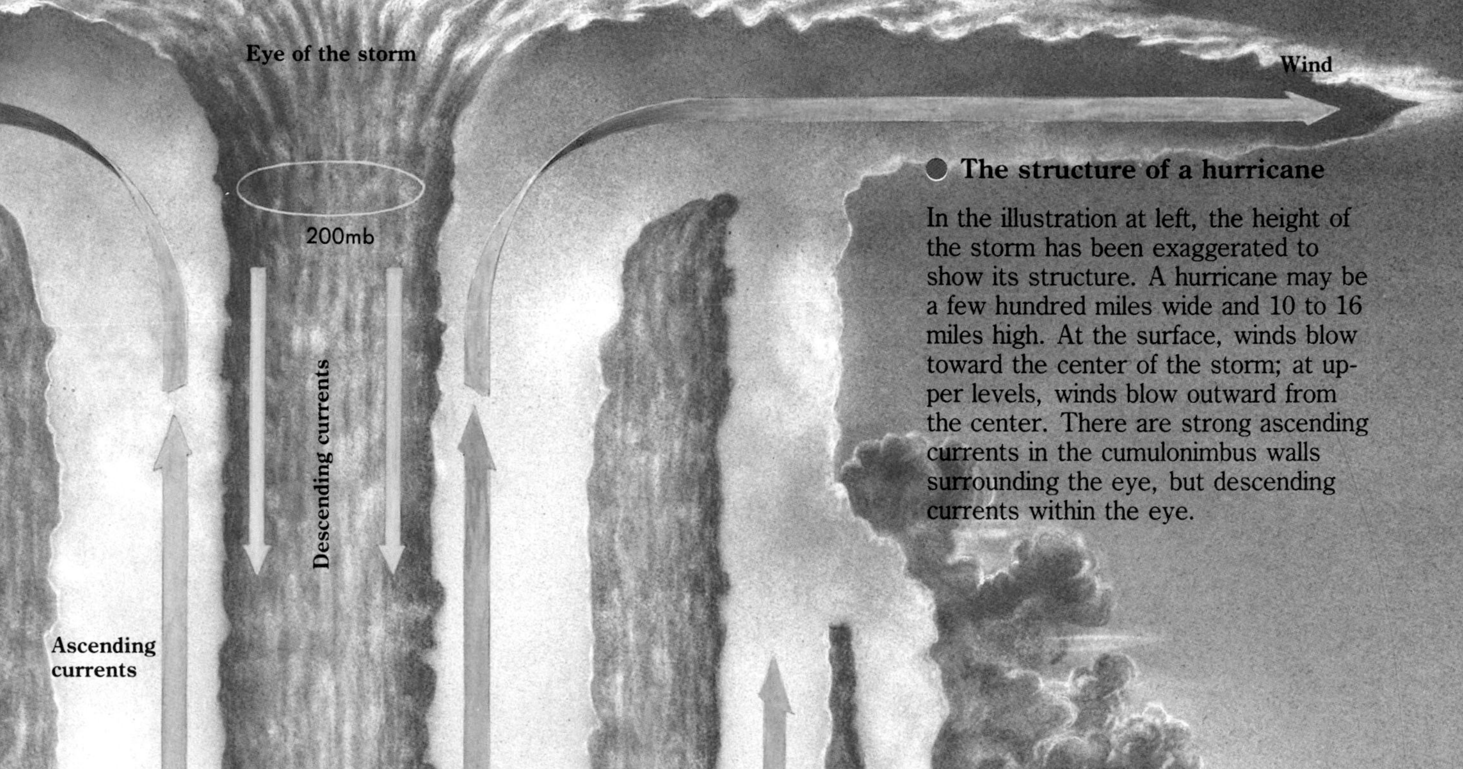


### ● The structure of a hurricane

In the illustration at left, the height of the storm has been exaggerated to show its structure. A hurricane may be a few hundred miles wide and 10 to 16 miles high. At the surface, winds blow toward the center of the storm; at upper levels, winds blow outward from the center. There are strong ascending currents in the cumulonimbus walls surrounding the eye, but descending currents within the eye.

◀ The aircraft flies in figure-eight patterns as it takes data in the eye of the storm.

▲ As the plane approaches the center, its crew uses radar to find the eye of the storm.

◀ A dropsonde collects data from the storm and the surface of the ocean.

# 5
# Aerial Wonders

The journey of sunlight and moonlight through the atmosphere creates an endless display of light and color. Some spectacles are taken for granted, like the blue of the sky *(page 22)* and the twinkling of the stars. But others are so strange or so glorious that one can only look at them in wonder.

Since time began, nature's fireworks have pleased and puzzled all who saw them. Early people tried to explain and understand the global light show through myth and superstition. The rainbow is commonly regarded as a sign of good fortune, while the glowing spheres produced by ball lightning were held by some to be the departing souls of the dead. But over the centuries, scientists began to discover the secrets of these mysterious displays and found them to be the products of very ordinary substances. Rainbows, for example, arise when sunlight strikes water droplets suspended in the air. Halos—those bright circles that sometimes appear around the sun and moon in cold weather—form through the action of light on suspended ice crystals. And mirages, often considered the products of fevered minds, need only light and unevenly heated air to develop.

By studying these atmospheric phenomena, scientists learn a great deal about the laws of physics and the atmosphere itself. But even as these dramatic displays are reduced to basic physical principles, they lose none of the beauty that amazed those who first saw them centuries ago.

A vivid rainbow arches over the pounding waters of Niagara Falls in western New York. Another rainbow, its colors pale and reversed in order, is faintly visible above the first. The rainbows formed when light struck water droplets in the falls' spray.

# What Is a Rainbow?

Rainbows owe their colorful arcs to the interplay of sunlight and moisture suspended in the air. Sunlight appears white to the naked eye but is actually made up of seven colors—red, orange, yellow, green, blue, indigo, and violet. When a sunbeam strikes the surface of a water droplet, such as a raindrop, the light refracts, or bends, and separates into its component colors. These colored light waves reflect off the inner rear surface of the raindrop, then bend again as they exit. Most of the time, light reflects off the droplets only once, producing a single rainbow. But sometimes light bounces inside the drops and reflects twice. This action results in the formation of two rainbows: a primary rainbow that is produced by the first reflection of light and a secondary rainbow above the first, produced by the second reflection.

Sunlight

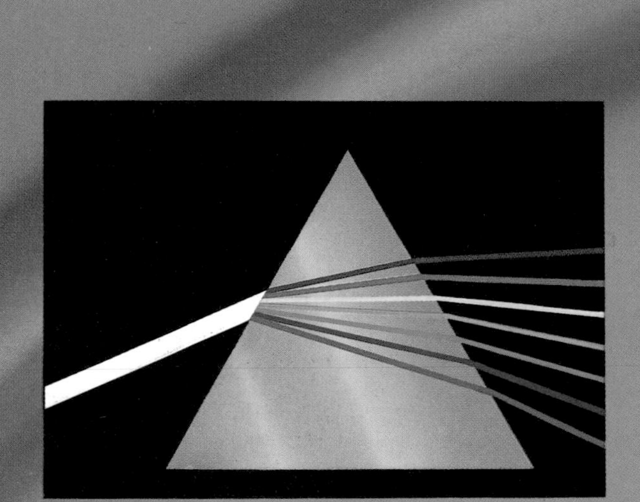

**Raindrops act like tiny prisms** in the atmosphere. As a ray of light passes through a prism *(above),* it separates into colored waves. Each wave bends to a different degree depending on its wavelength. Red light has the longest wavelength and bends the least, while violet light, with the shortest wavelength, bends the most.

40°

42°

**The legacy of a rainstorm,** a rainbow rises from a field in Port Alsworth, Alaska.

Red light reflected

Violet light reflected

## Creating a rainbow

Some of the sunlight entering a raindrop passes right through *(right).* Other rays of light are bent away from an observer on the ground. Only those rays that reflect between 40° and 42° *(left)* will show up as a rainbow to an observer on the ground. All droplets reflect the red waves of sunlight at an angle of 42°, so an observer will see red light only from droplets suspended at that altitude. Violet light, reflected at an angle of 40°, is visible only from droplets suspended 40° above the horizon. The other colors are reflected at angles between those of red and violet, which is why red is always the top color in a primary rainbow, violet is the bottom color, and the other colors appear in between.

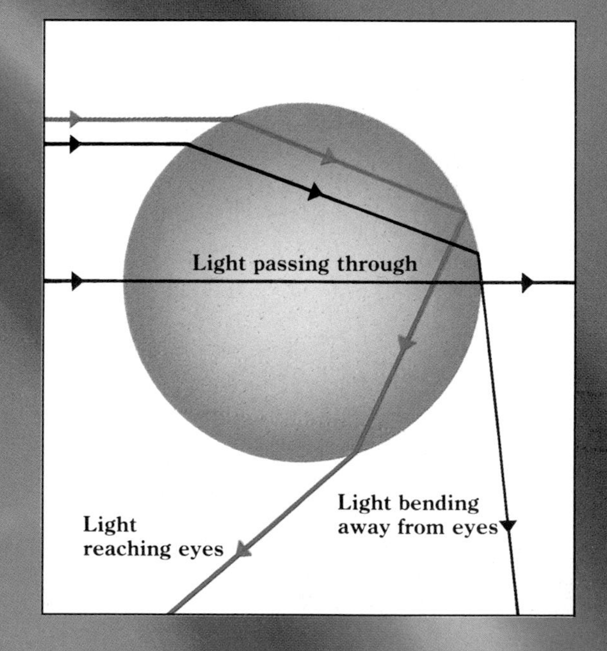

# What Are Mirages?

Mirages are tricks nature plays on the eyes. Common mirages are an oasis in the middle of a burning hot desert, a pool of water on a dry paved road that seems to reflect the image of a car, or a ship sailing across the sky above an ocean. These images may appear blurred, unusually large, or even inverted (upside down); sometimes two or more identical images may appear.

Such optical illusions are created when the speed and path of light rays traveling from distant objects are altered by atmospheric conditions. Normally, light passes through the atmosphere undisturbed, and an observer looking at a faraway object sees it as it really is. But sometimes, when there is a great difference in air density—or temperature—from one layer of air to another, the atmosphere begins to act as a lens, bending light rays and creating distorted images.

**When the air near the ground** is very hot, as in a desert, objects sometimes appear to float above the surface.

**Cars on a hot highway** seem to be reflected in a shimmering pool of water. The pool is really an inverted image of the sky caused by bending light.

### How a floating mirage forms

When surface air temperature is very low, as over frigid oceans or lakes, one or more images of a distant object—such as the boat shown below—may appear in the sky above the object. Called superior mirages, these images occur when light rays *(red lines)* rise from a cold air layer, strike warmer air, and bend downward toward an observer's eye. Sometimes multiple images occur. They may be upright or inverted, depending on the object's distance and the temperature of the layers of air.

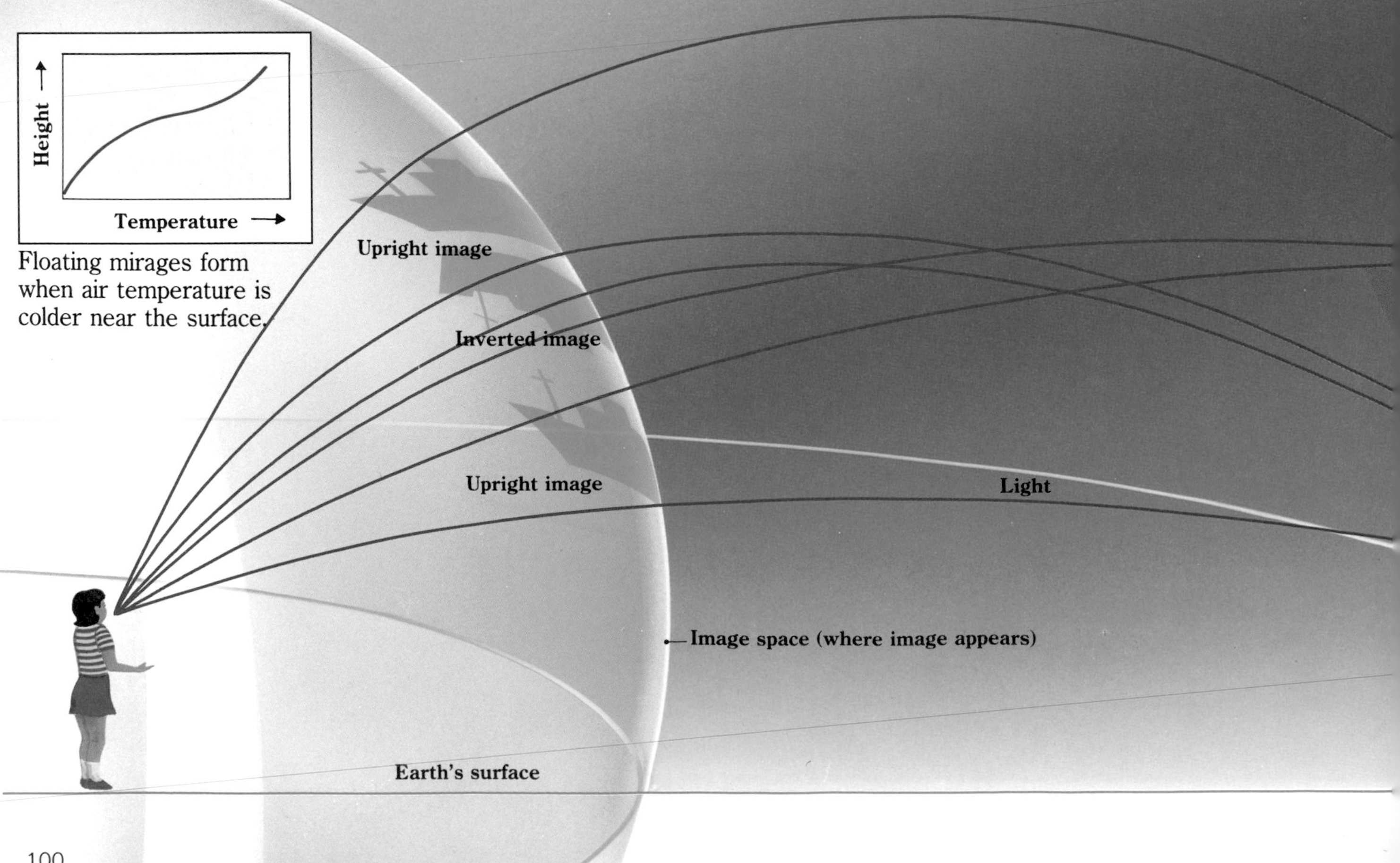


Floating mirages form when air temperature is colder near the surface.

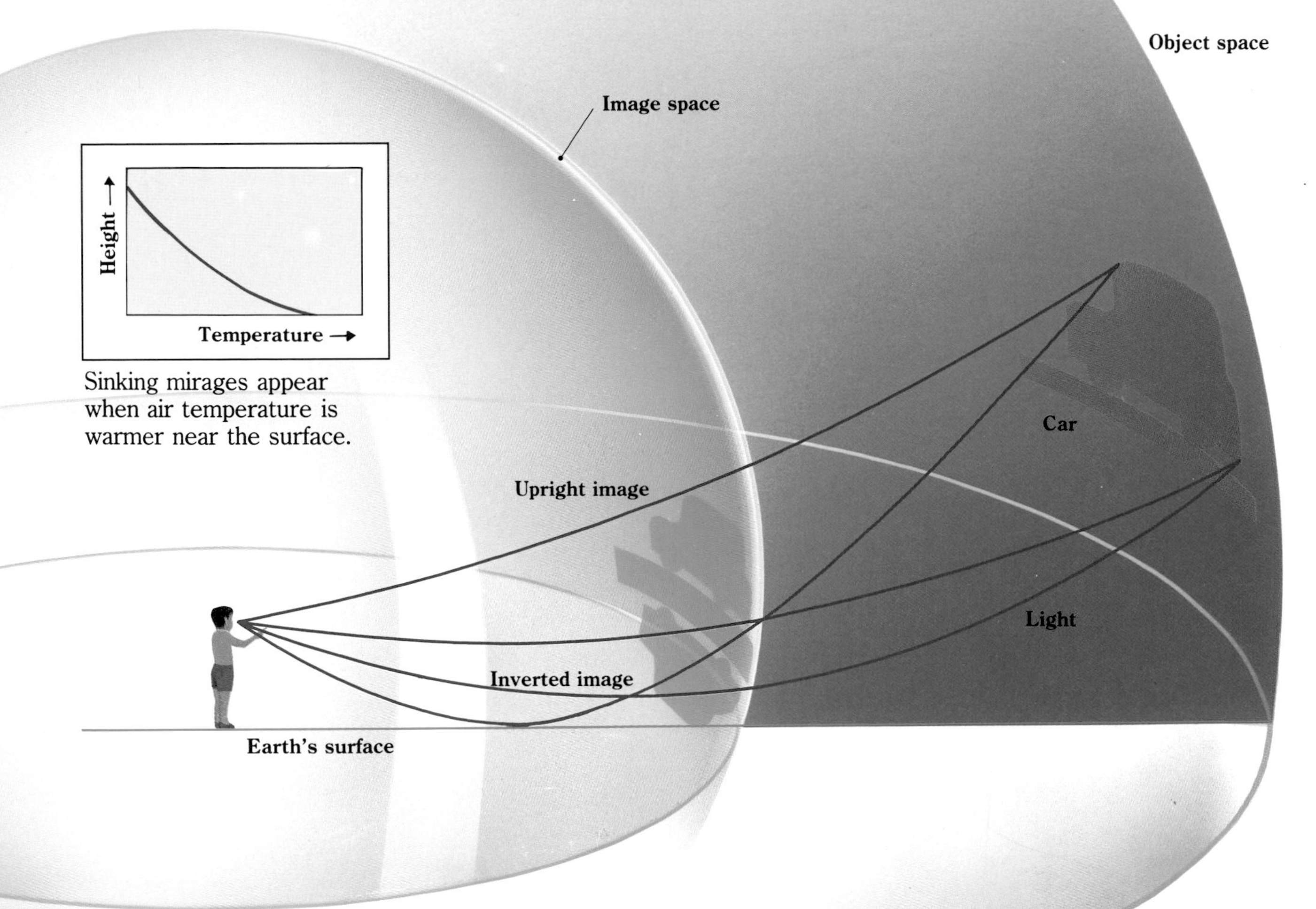


Sinking mirages appear when air temperature is warmer near the surface.

## How a sinking mirage forms

Light rays always bend with the cooler air on the inside of the curve. When the ground is very hot, light rays traveling toward the ground from the top of an object—such as the car above—will bend up from the hot surface air toward cooler air at eye level. Light traveling from the bottom of the object—say, the car's wheels—will bend upward at a smaller angle. The result is a two-image inferior mirage: The object appears once in its normal position and again upside down (because some of the light rays crisscross).

### Heat, light, and mirages

The kind of mirage seen depends on the temperatures of different air layers. Four mirages are shown below; a downward arrow indicates an inverted image.

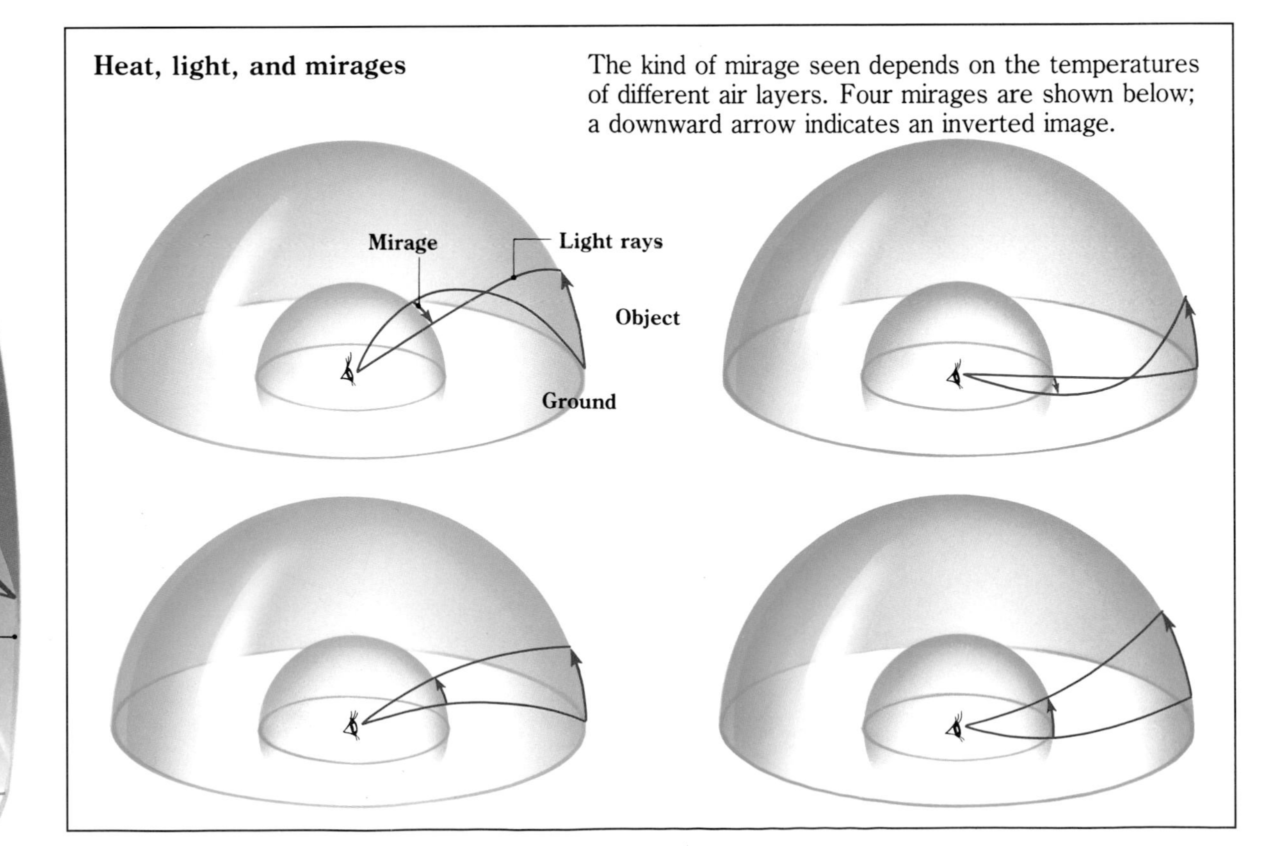


Ship

Object space (space occupied by object)

# How Do Solar and Lunar Halos Form?

When the weather is cold, bright rings occasionally appear around the sun or moon. These rings, or halos, form when light strikes certain ice crystals in the atmosphere, and those crystals bend and focus the light in a particular way. The most common halo is a 22° halo, or inner halo; the ring it forms around the sun or moon has a 22° arc (measured as though a line drawn from sun to halo formed the base of a triangle and the observer stood at the peak, as shown below). Less common is a 46° halo, also called an outer halo; it has a 46° arc and thus appears as a larger circle around the sun or moon.

Crystals also reflect light in other ways, producing such spectacular phenomena as multiple images of the sun called parhelia, or mock suns; vertical beams of light called solar pillars, which can appear above the rising or setting sun; and arcs above and below the sun called Parry arcs.

**Inner halos** *(above)* can be brightly colored; outer halos are often white.

**Images formed by ice crystals**

The diagram below shows some of the dazzling displays ice crystals can create by bending, focusing, or reflecting light. Many of these same images also may appear around the moon.

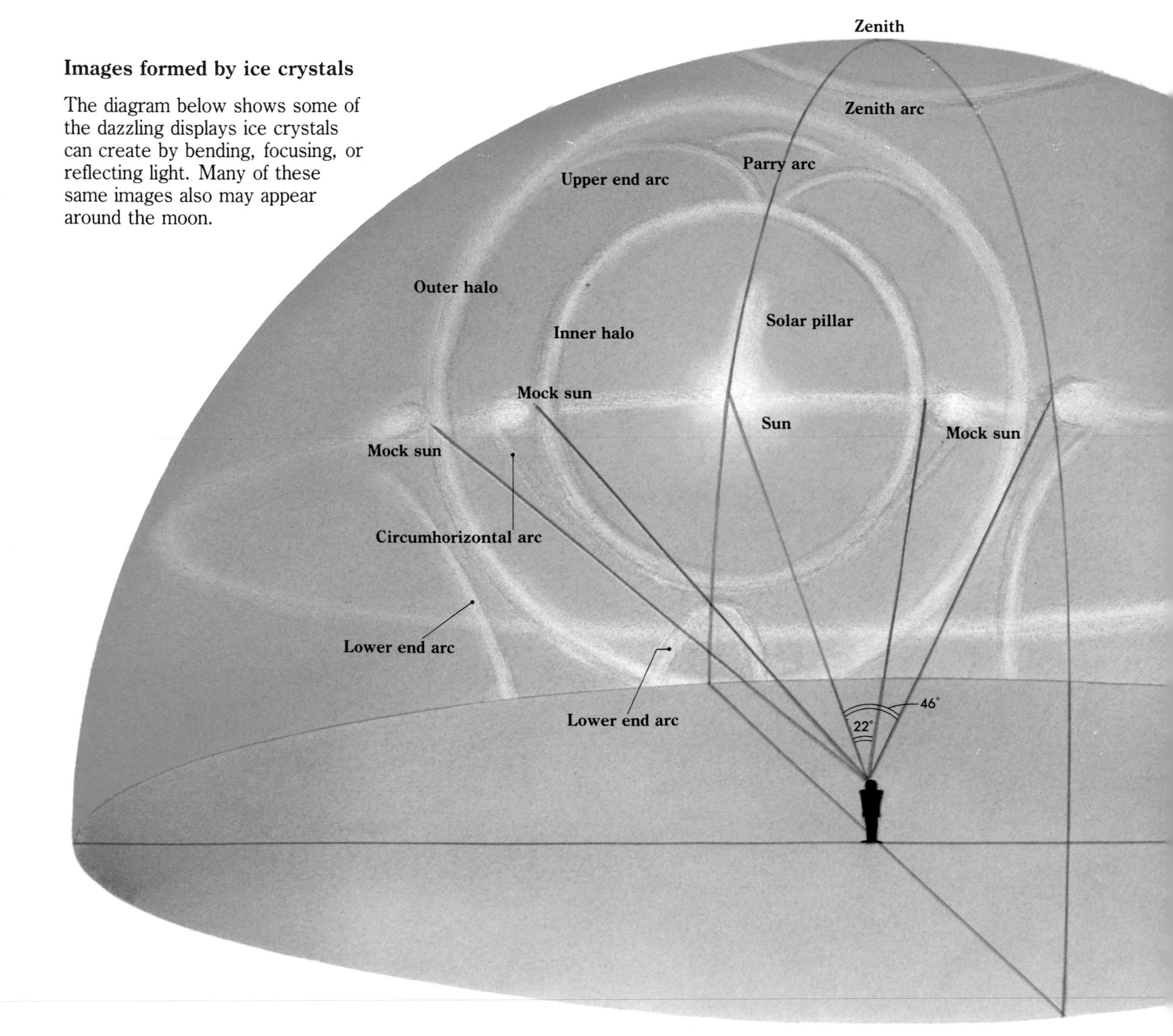

## Halos of crystal and light

Halos are formed by hexagonal (six-sided) crystals. When light enters the end of one of these crystals, it forms an outer halo that from the ground appears at a 46° arc from the sun or moon. When light enters from the side, however, it forms an inner halo at a 22° angle from the sun or moon.

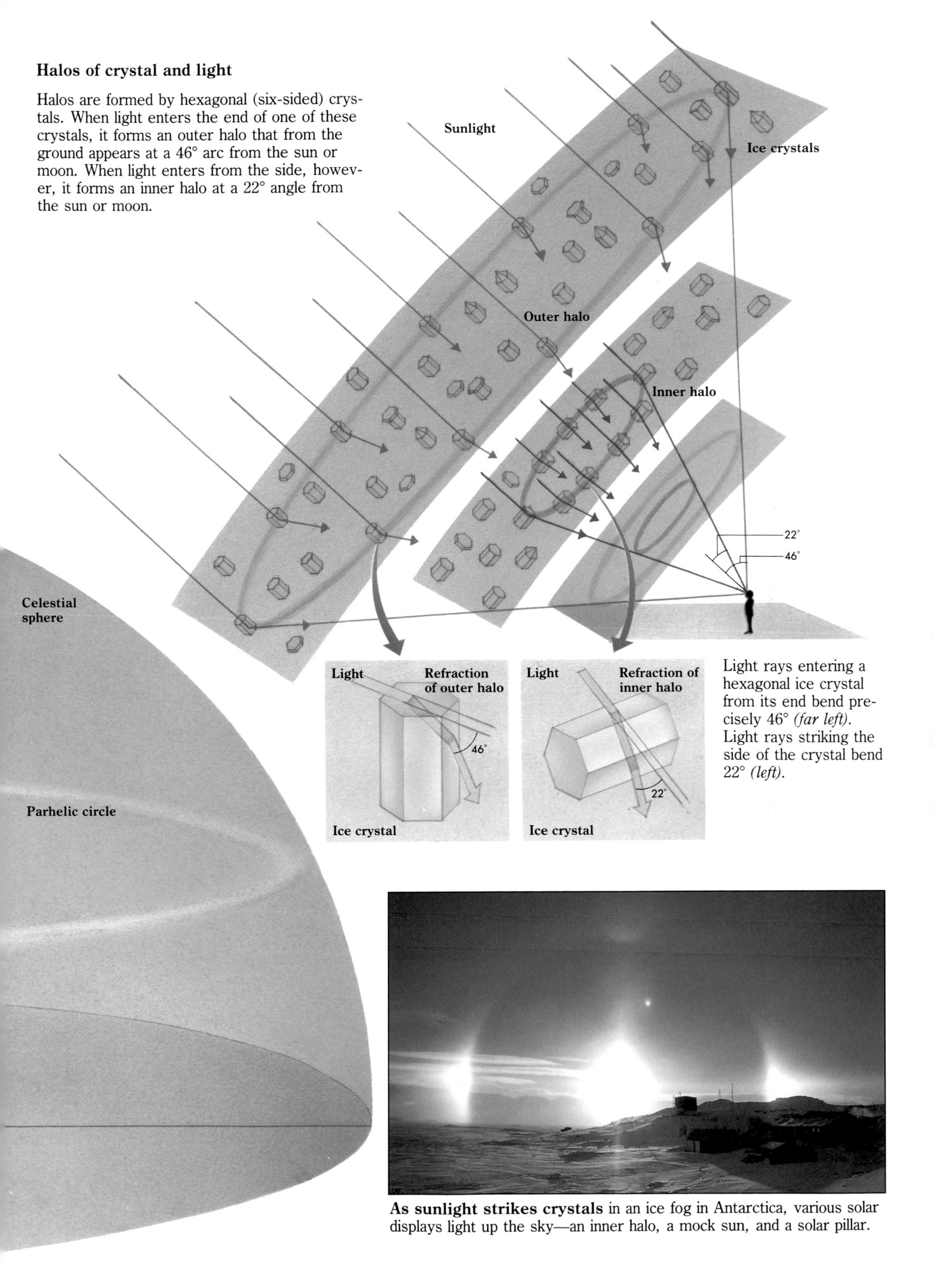


Light rays entering a hexagonal ice crystal from its end bend precisely 46° *(far left).* Light rays striking the side of the crystal bend 22° *(left).*

**As sunlight strikes crystals** in an ice fog in Antarctica, various solar displays light up the sky—an inner halo, a mock sun, and a solar pillar.

# What Causes Mountain Halos?

Often observed by mountain climbers, mountain halos occur at high altitudes when the sun shines from directly behind an observer into a fog or cloud bank directly ahead. Tiny water droplets in the fog or clouds reflect the sunlight, forming the rainbowlike halos; because the sun is behind the observer, a shadow appears at the center of the halo.

To create the multiple, concentric circles another step must occur—diffraction, the curving of light around an obstacle. As the rays split around the observer, they curve slightly. Some rays recombine but others do not, creating a series of alternating bright and dark circles called a diffraction pattern *(below)*. The colored halo forms within the bright bands of this pattern.

**The elongated shadow** of an observer appears in the center of this mountain halo.

**The principle of diffraction**

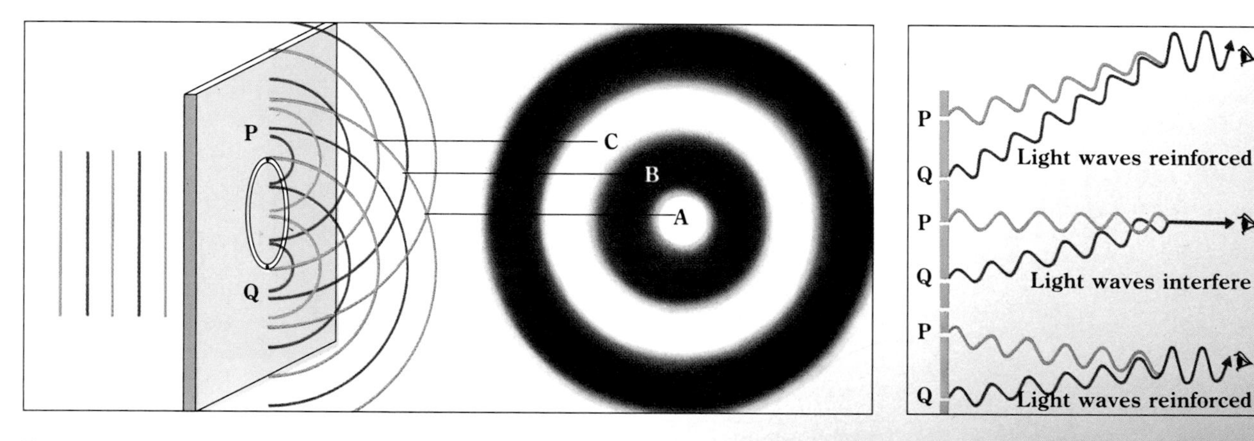


To understand the diffraction of sunlight, imagine an observer as a circle cut out of a panel *(above, left)*. As parallel beams of light *(dark and light lines)* pass through the circle, they curve outward. At points A and C, the waves meet again, reinforce each other, and become especially bright. At point B, the waves interfere, creating a dark region. Although the above diagram only shows the diffraction pattern generated at points P and Q, diffraction occurs at all points around the circle—or observer—generating the circular diffraction pattern *(above, right)* characteristic of mountain halos.

Diffracted light waves that recombine, as with A in the above diagram, create light that is twice as bright as each component wave. If the waves interfere, as in B, they cancel each other out and the observer sees no light.

## Coronas

Coronas are bright bands of light sometimes seen around the sun and moon. Light passing from the sky through suspended water droplets is curved, or diffracted, by the edges of the droplets. The disturbed light waves then combine to produce concentric bands of color *(right)*.

A lunar corona *(left)* blazes as water droplets in the clouds diffract the moonlight.

## Making a mountain halo

An observer can see a mountain halo—and his or her shadow at its center—when the sun shines from directly behind the person into clouds or fog directly ahead. The halo's circles appear when the incoming light is diffracted, its waves curving around the observer and reflecting off the clouds. The light waves then recombine in a circular diffraction pattern of light and dark bands; each wavelength of light is a different color.

# What Is Ball Lightning?

Ball lightning is one of the most bizarre of all atmospheric phenomena. While ordinary lightning comes down in bolts from the sky, ball lightning is a fireball with a tail that moves close to the ground. It typically follows telephone or power lines for a few seconds, then vanishes. According to some reports, fireballs also enter homes. Slipping through doors or windows, or diving down chimneys, they dart about furiously, then make a quick exit, leaving scorch marks in their wakes.

People have reported seeing fireballs for centuries. But because ball lightning is so rare—and so fleeting when it occurs—it was not until the mid-twentieth century that scientists began studying it seriously. No true explanation of ball lightning yet exists, but physicists have some promising theories, a few of which are illustrated here.

**Trailing a fiery tail,** glowing ball lightning crosses railroad tracks in Japan in this 1987 photograph.

### Features of ball lightning

Blazing with the brightness of several hundred light bulbs, ball lightning ranges in size from 6 inches to 2 feet in diameter and in shape from spherical to oval to wispy. Its color is usually white, sometimes with an orange or blue cast. But as varied as fireballs are in their appearance, they differ much more widely in behavior. Following paths that are straight or wildly curved, fireballs can float lazily over the ground for minutes or race by at thousands of miles per hour.

### Occurrences

Fireballs tend to occur in places where the air is stagnant, as over marshes and valleys. In cities, they usually form near high-tension wires, telephone lines, and the corners of metal buildings and towers. Nearly half of all reported fireballs appear inside homes and buildings, with the fireball usually entering through a door or window.

## How ball lightning forms

Three theories relate fireballs to lightning strikes. In the electricity theory, lightning strips hydrogen atoms from airborne water molecules. The hydrogen bonds with molecules containing carbon to form a ball of hydrocarbon molecules that emit light as they release excess energy. In the aerosol plasma theory, electrically charged particles called aerosols form a sphere. When lightning strikes, the sphere becomes energized and glows. In the electromagnetic wave theory, static electricity builds up in clouds and trees, generating electromagnetic waves that bounce off the ground. Where the waves meet, electromagnetic energy excites the air, forming a lightning ball.

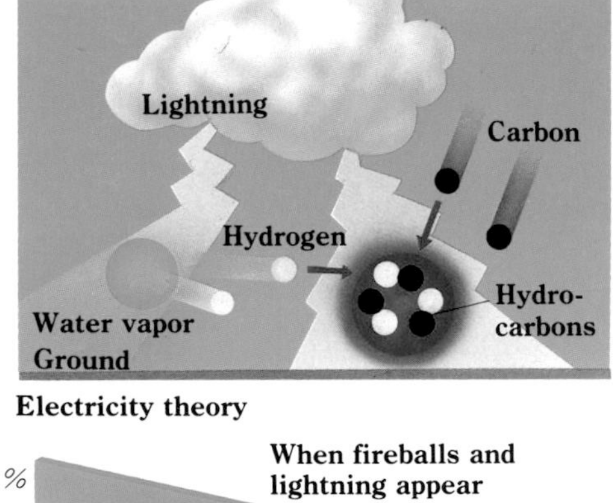

Electricity theory

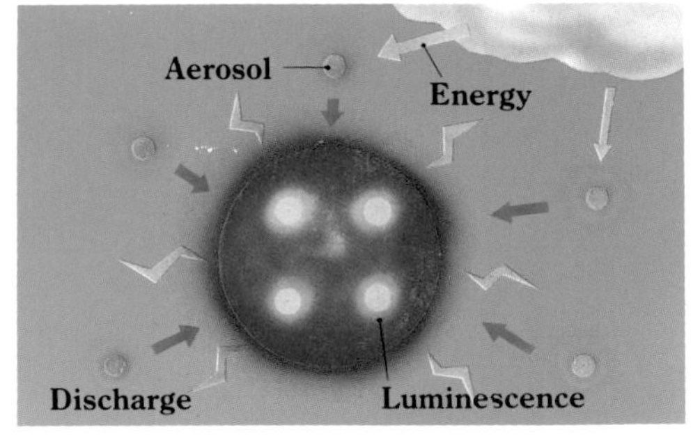

Aerosol plasma theory

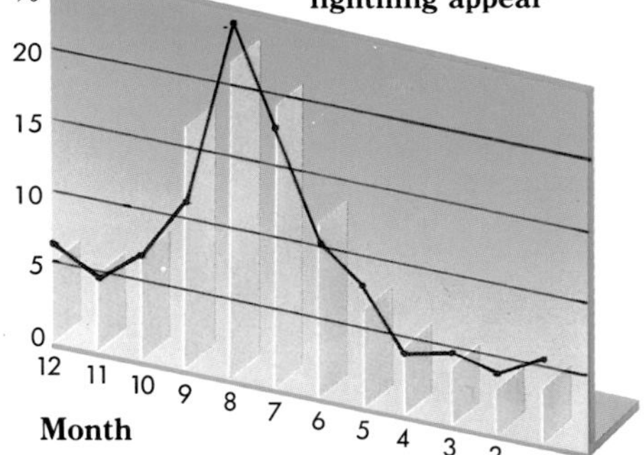


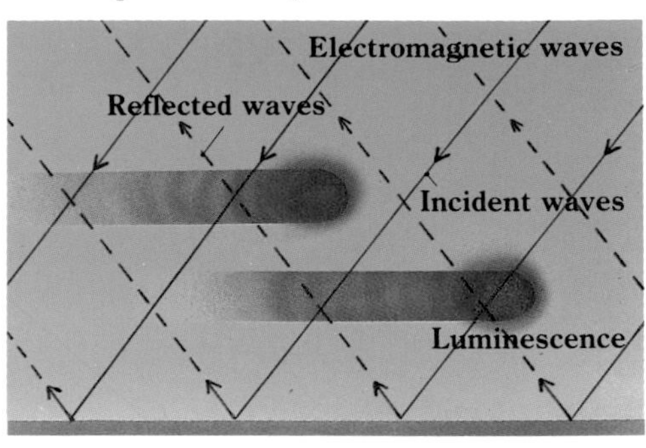

Electromagnetic wave theory

### In the lab

In an experiment to find out the true nature of ball lightning, researchers developed a device *(below)* that, through an antenna, gave off electromagnetic waves in an ethane-filled room, producing a lightning ball. The apparatus was based on the theory *(left)* that electromagnetic waves from lightning may ultimately come together at a point where their collective energy strips electrons from molecules of air. These charged molecules then begin to glow, producing a fireball.

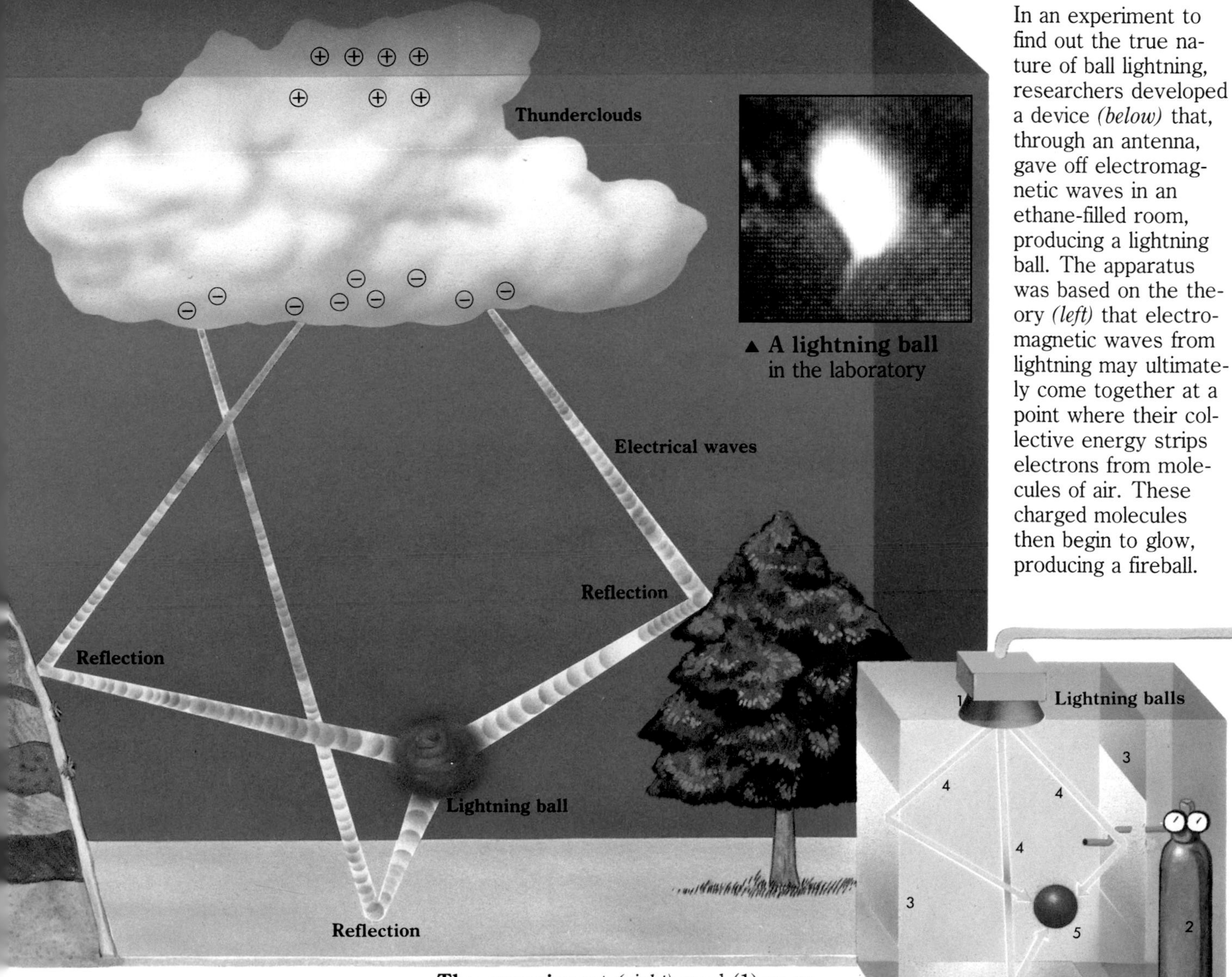


▲ **A lightning ball** in the laboratory

**The experiment** *(right)* used (1) an antenna, (2) ethane, (3) reflectors, and (4) electric waves, making (5) a lightning ball.

# 6

# Watching the Weather

In the days many centuries ago when weather was seen as an unpredictable act of God, people relied on folklore and their own knowledge of the clouds and winds to try to forecast the weather. Then, in the seventeenth century, the mercury barometer was invented, joined in the eighteenth century by the Fahrenheit and Celsius thermometers. Weather observation stations began to spring up, and reliable forecasts began to aid the average citizen.

By the end of the twentieth century, about 10,000 weather reporting stations were established around the world. Every few hours, these stations collect information on temperature, pressure, precipitation, and the like, compiling the data into maps. These maps, in turn, help scientists prepare weather forecasts for that day, the next day, and a few days, weeks, and months in the future. The further into the future the forecast must reach, the more difficult it is to prepare accurately.

At the most sophisticated weather agencies and research centers, scientists now use powerful supercomputers to create models of the weather. Millions of observations of the Earth as a whole may be fitted into a model that includes equations drawn from basic laws of physics. Even on a supercomputer, a prediction made from such a model can take more than 100 hours to run. In time, faster and more accurate computer models may revolutionize forecasting.

Weather observers hike to a mountaintop weather station topped by the characteristic white dome of a weather radar antenna. The observers will work a three-week shift before coming back down.

# How Do Satellites Keep Track of Weather?

Weather satellites see the Earth in many different ways. Far above the planet's surface, their instruments capture visible, infrared, and microwave radiation from the Earth below. In addition to providing visual images, satellites are able to sense temperatures of the ground, the sea surface, and various levels of the atmosphere; they also measure winds above the oceans and atmospheric moisture.

Two kinds of satellites watch the weather from outer space. Geostationary satellites, positioned 22,300 miles above the equator, circle the Earth at exactly the rate that the Earth turns so they stay over the same place on the surface. This allows them to carry out continuous observations over a wide area. Polar orbit satellites travel in north-south orbits, so the Earth turns under them. They can observe most of the Earth, including the polar regions that geostationary satellites cannot see.

**Ground station monitors** show satellite images.

### A satellite transmits data

A satellite sends raw data *(blue)* to a communication center that forwards it to weather stations. The satellite also relays processed images *(orange)* from the ground to remote stations and collects data *(purple)* from remote buoys, ships, aircraft, and island observatories.

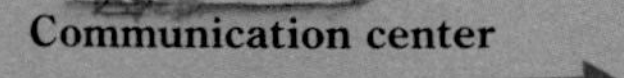

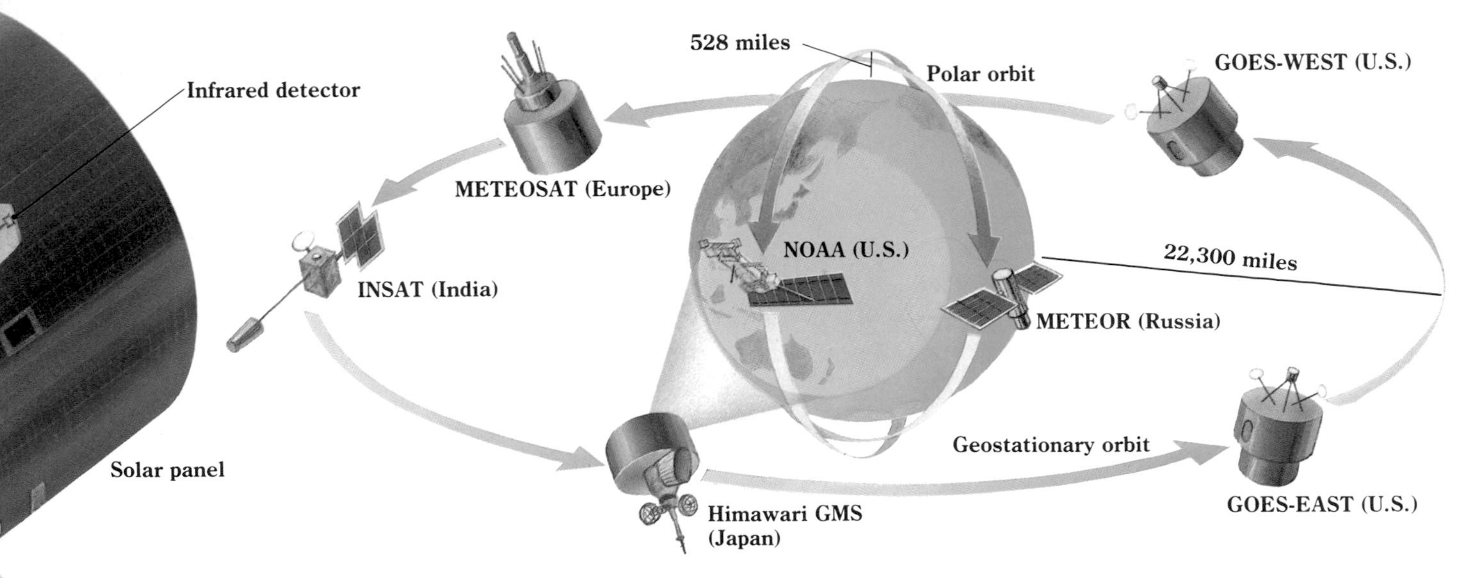


### A satellite tracks weather

Geostationary satellite instruments scan the Earth's surface in narrow bands *(left)* to build up images. After each west-to-east scan, the satellite's instrument mirror changes its angle slightly until the satellite has a complete image of the surface below.

### A global satellite network

In the late 1980s, five geostationary weather satellites girdled the Earth, in addition to several polar orbit satellites *(above)*. Between them, the satellites, operated by several nations, maintained constant surveillance of the world's weather.

**Clouds swirl** around low-pressure regions in the Pacific Ocean in a satellite image. Computer processing adds colors.

Observation data

Observation ship

East

▲ **Scanners aboard satellites** monitor small areas of the Earth's surface continuously. The infrared detector sees sections 4 miles square; the visible-light detector sees half-mile squares.

# Why Watch the Upper Atmosphere?

Most of the weather on Earth occurs in the troposphere, the lowest layer of the atmosphere. However, the troposphere is influenced by the layers above it: the stratosphere, ranging from 10 to 30 miles above the surface, and the mesosphere, from 30 to 50 miles up. For this reason, scientists study the upper atmosphere in order to understand the region below.

The stratosphere, for instance, sometimes experiences sudden heat spells in which the temperature increases by dozens of degrees for several days. Every two years or so, the winds in the stratosphere above the equator reverse themselves, blowing from the east and then from the west again.

The weather in the stratosphere is also important in its own right because jet aircraft fly there. Recently a new reason for understanding the upper atmosphere has emerged: The stratosphere is the location of the ozone layer, which protects Earth's surface against damaging ultraviolet radiation from space.

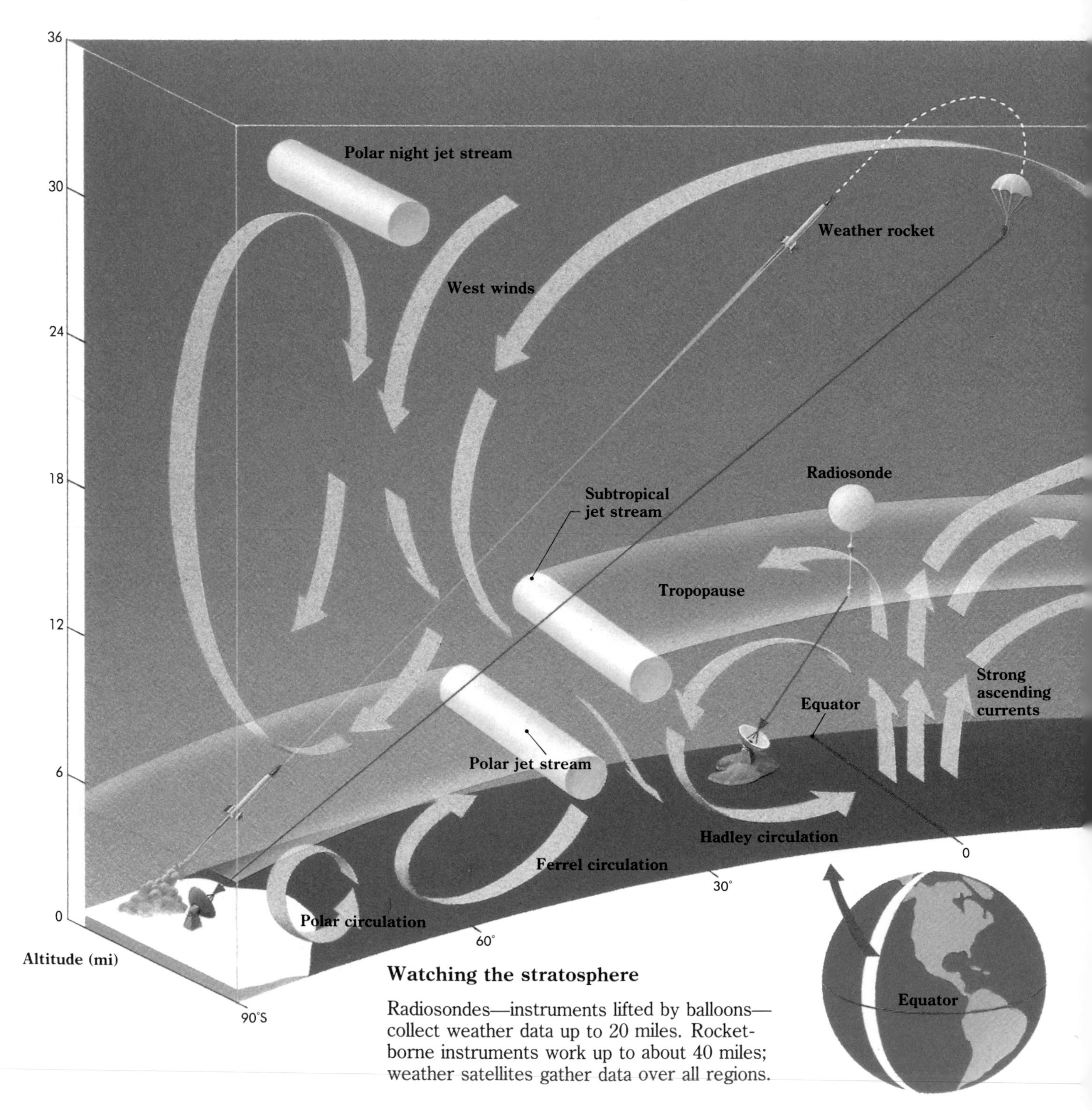


**Watching the stratosphere**

Radiosondes—instruments lifted by balloons—collect weather data up to 20 miles. Rocket-borne instruments work up to about 40 miles; weather satellites gather data over all regions.

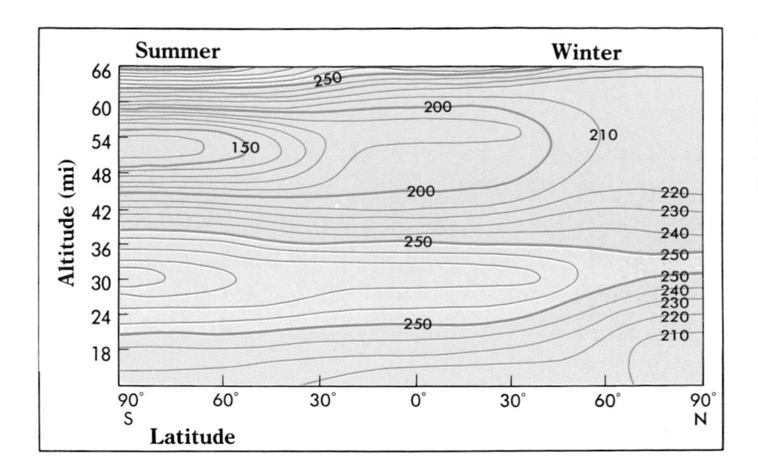

## Temperatures aloft

Atmospheric temperatures (in degrees Kelvin) change with latitude and altitude. For instance, in the Southern Hemisphere summer, warmer regions surround a cold area 54 miles up.

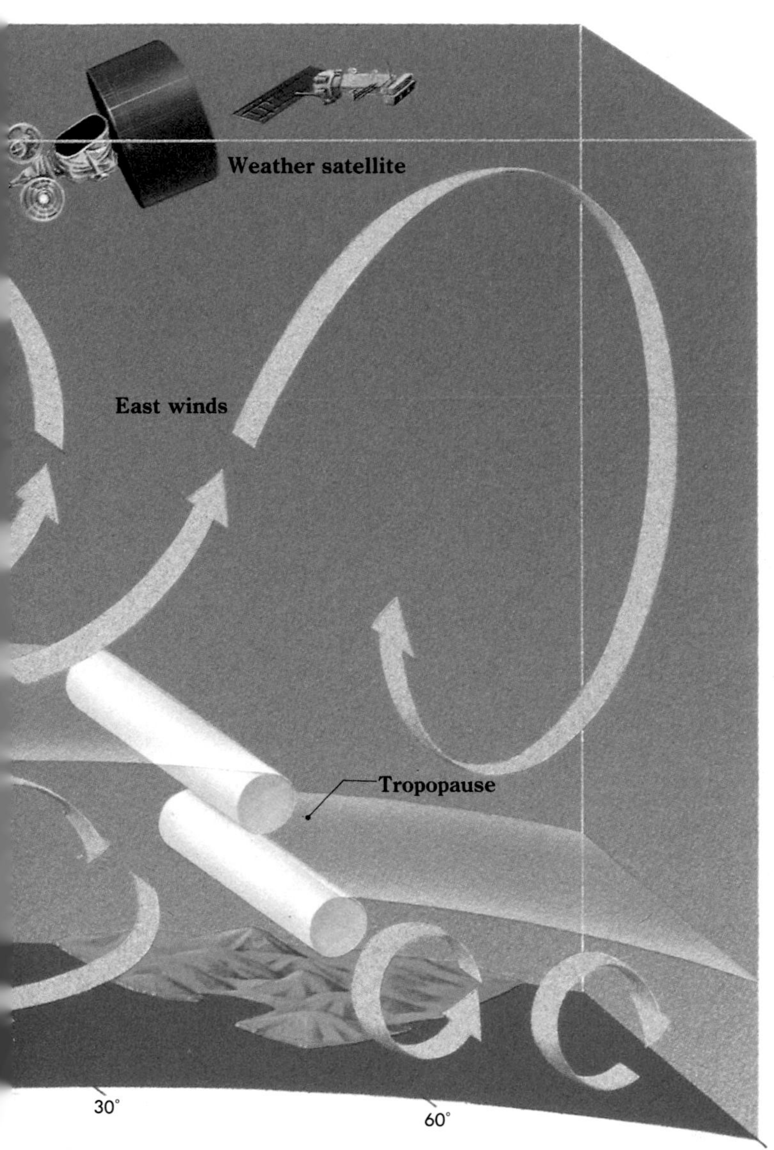


## Winds in the stratosphere

Air moves in the stratosphere *(above)* because the ozone layer is heated by ultraviolet rays. Starting from the summer hemisphere, where the heating of the ozone is greater, air moves to the winter hemisphere, where it is less. The Earth's spin bends the movement into west winds in winter and east winds in summer. At high latitudes, the troposphere also sees a small opposite circulation. Air rising from the troposphere to the stratosphere divides into north and south currents.

## Heat, cold, and altitude

Maps of temperatures (in degrees Kelvin) at different altitudes on a February day *(below)* show a complex pattern of warmer and colder air at every level. In general, temperature in the stratosphere increases with altitude, while in the mesosphere it decreases.

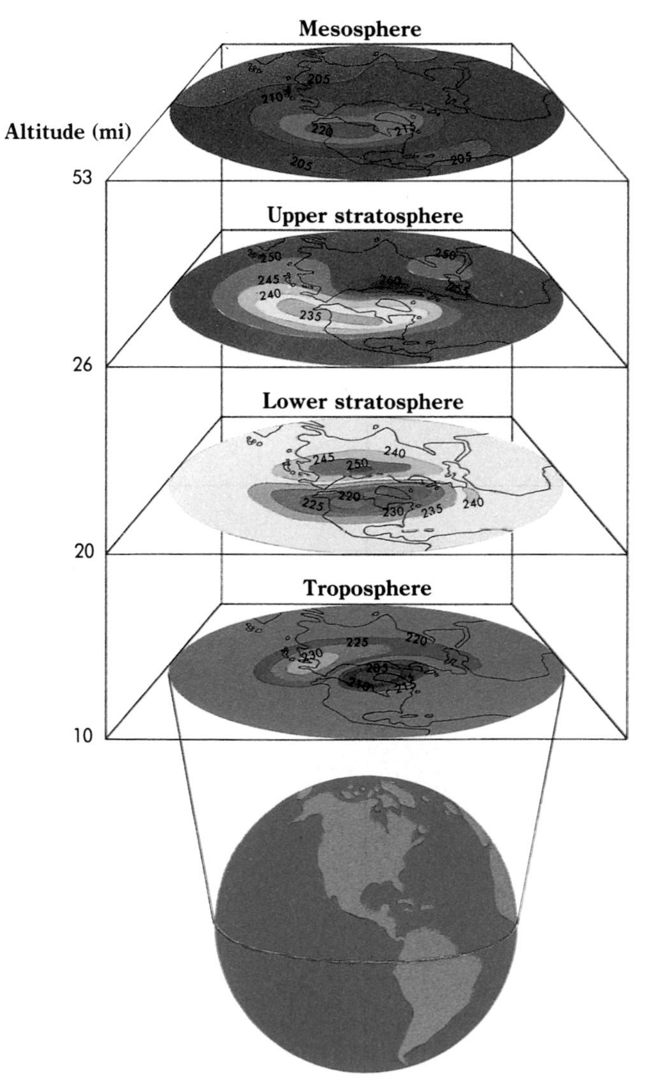

# How Does Weather Radar Work?

Weather radar uses radio waves to collect information about precipitation such as rain and snow. When radio waves from a radar antenna hit particles of precipitation in the air, they bounce back to the antenna. A computer gauges the distance to the precipitation by measuring the time a radar pulse takes to return. The radar also detects the direction of the precipitation and its intensity, as shown by the strength of the return echo.

The distance that weather radar can see depends partly on the strength of its beam and partly on the altitude of the antenna. Radio waves, moving in a straight line, rise from the curved surface of the Earth. At a range of 180 miles, the waves from a radar at sea level are more than 4 miles high, above most precipitation.

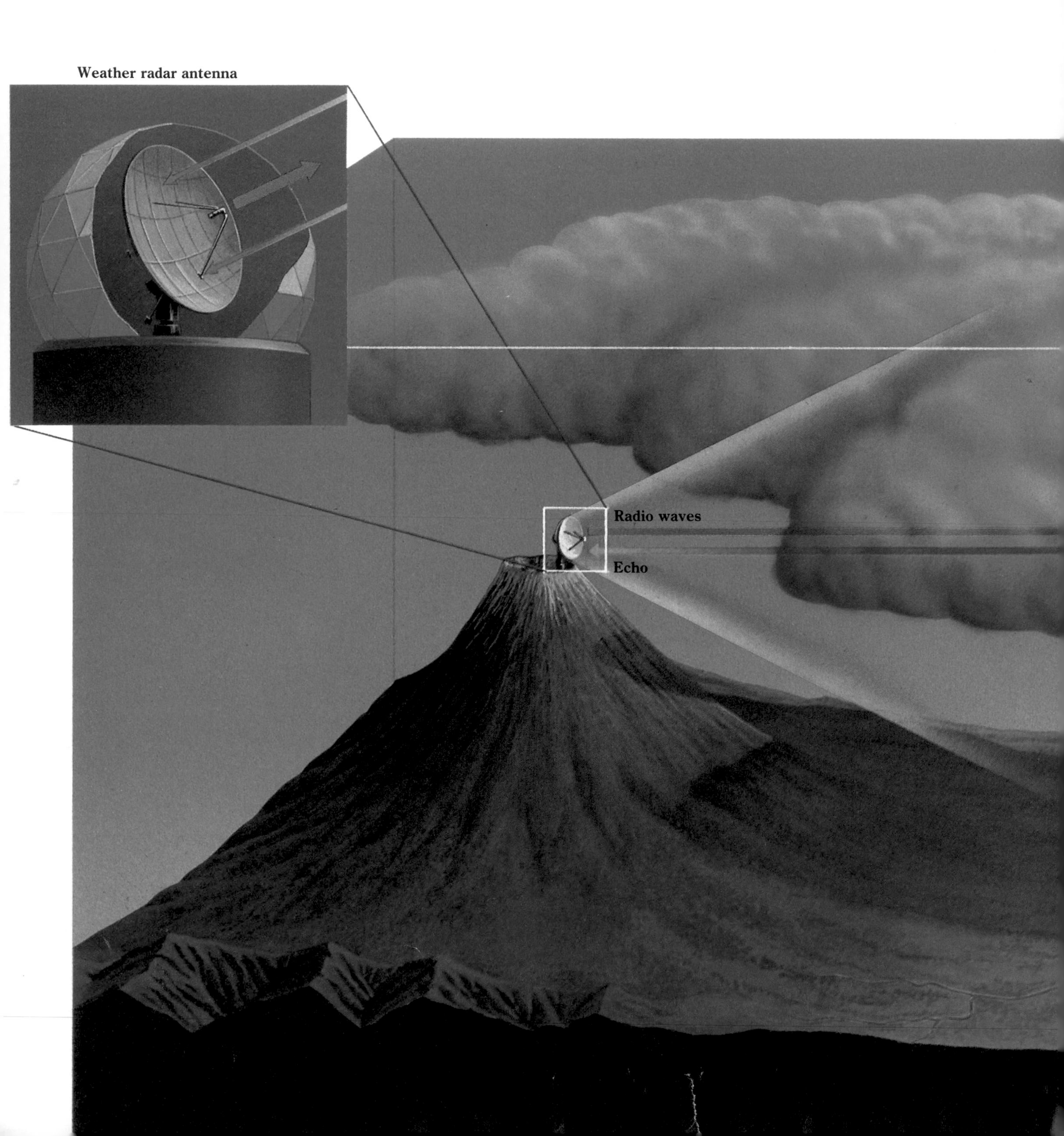

## Range of weather radar

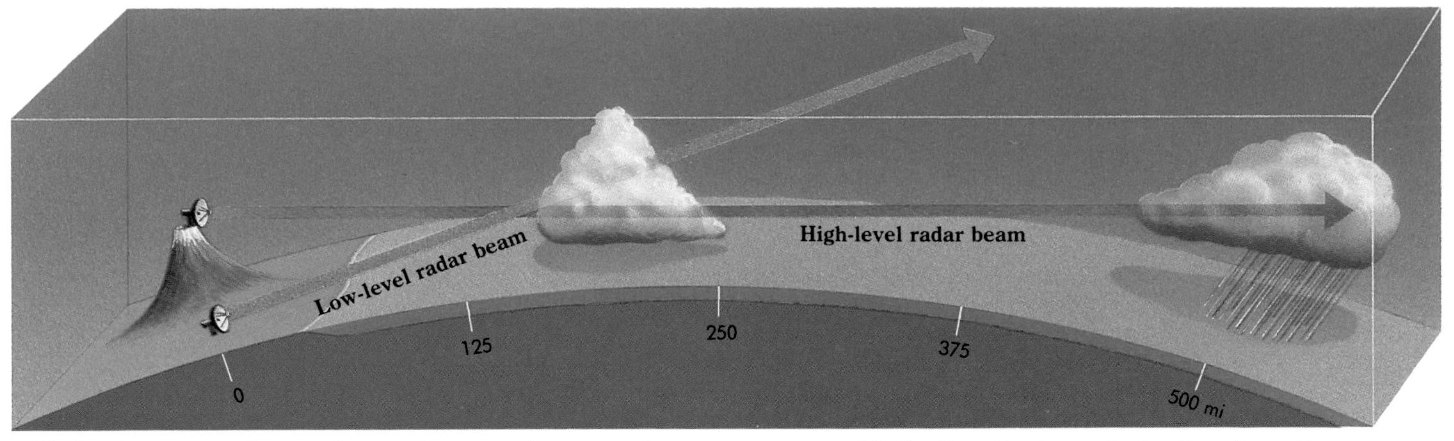


Radar from a weather station on the ground must angle its beams away from the Earth's surface. A mountaintop radar, however, can emit its beam at a slight downward angle; a beam that starts out 12,400 feet above sea level crosses the horizon at a range of about 155 miles. At a range of 340 miles, the beam is 4 miles above the surface, the highest altitude of most precipitation-bearing clouds. Hurricanes and other powerful storms often rise to an altitude of 6 miles or higher, making them visible to a mountaintop radar at a range of 500 miles.

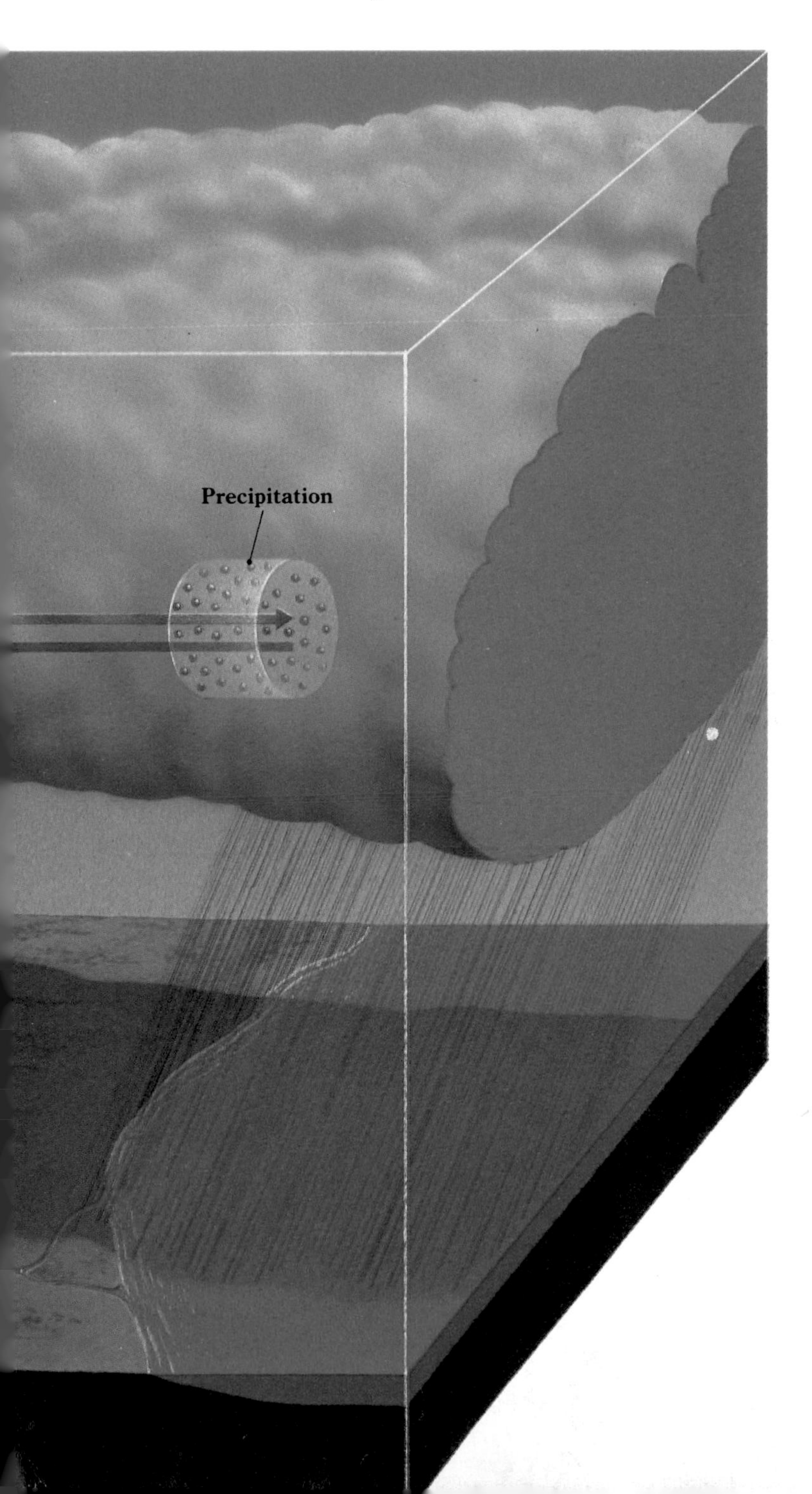

**The parabolic antenna** atop Japan's Mt. Fuji is 16 feet wide and protected by a dome. Its radio beam has a power of 1,500 kilowatts.

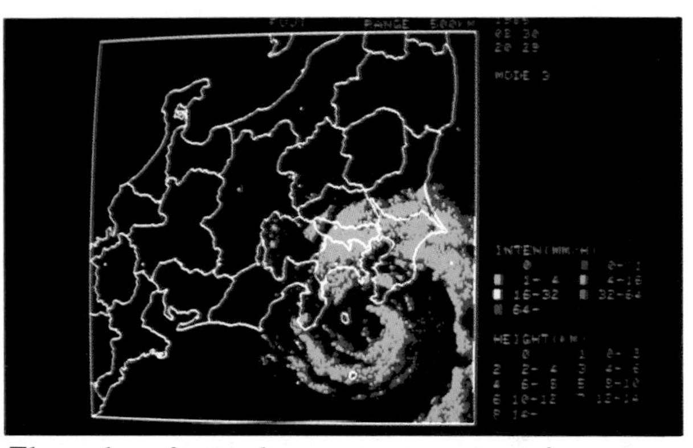

**The echo of a typhoon** appears on a radar screen, showing the clear eye of the storm at the center of a spiral of clouds.

### Radar and rain

Radar sends out high-frequency waves known as microwaves, with typical wavelengths of between .1 and 10 centimeters (up to about 4 inches). These waves will pass through clouds but bounce off raindrops or snowflakes *(left)*. The shortest radio waves produce a strong cloud echo but are easily absorbed by precipitation; longer waves are not absorbed but result in a weaker cloud echo. If the precipitation is moving toward or away from the radar, the returning signal's wavelength will change slightly, telling observers where weather systems are moving.

# Who Watches the Weather?

### Balloons and airplanes

Atmospheric drift balloons, which can stay aloft for days, return data about long-term weather trends; aircraft can be sent to areas of immediate interest, such as storm centers.

### Radiosondes and rockets

Radiosondes, sent up twice a day, chart conditions up to about 20 miles high; weather rockets, launched weekly, range as high as 40 miles.

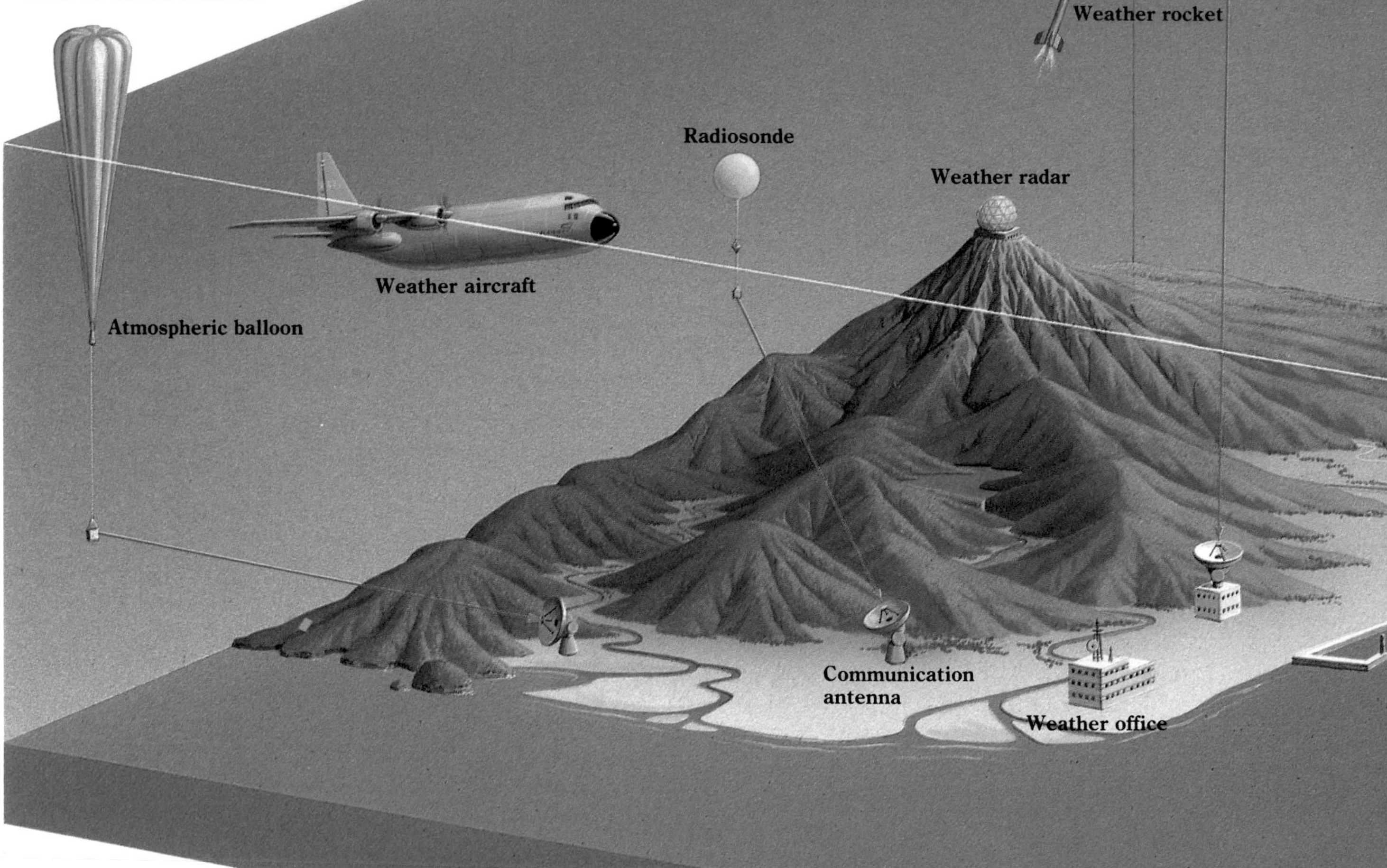


Reliable weather reports are based on detailed information from many different sources. Ground-based observations, carried out since ancient times, are now conducted by automated sensing devices as well as by humans who monitor the weather from thousands of observation stations. Many stations also launch radiosondes and weather rockets to check conditions in the upper atmosphere. Special weather ships and automatic weather buoys, as well as ordinary merchant ships, provide information on the weather at sea. National weather organizations collect most of these observations, but observation stations around the world exchange data to produce the most accurate, long-term weather reports and predictions.

**A weather rocket** thunders aloft from a Pacific weather station; its payload of instruments will radio back observations.

## Adding the view from space

Weather satellites in polar orbits and in geostationary orbits above the equator return information about cloud formations and ocean surface temperature around the world.

## ● Watching from the ground

At ground-based weather stations, observers record data once an hour. At automated weather stations, the recording devices update their readings every hour.

Polar orbit satellite

Automated weather station

Geostationary satellite

Weather ship

Weather station

Remote weather buoy

## Weather watch at sea

Weather observation ships record the same kind of readings as land-based stations. They also routinely monitor ocean temperature and salinity. Remote buoys relay weather and water data from critically important ocean locations.

## Worldwide weather network

Because conditions around the globe affect even local weather, accurate reports require observations from around the world. International weather stations exchange a great deal of information over special communication networks, including undersea cables and satellite links.

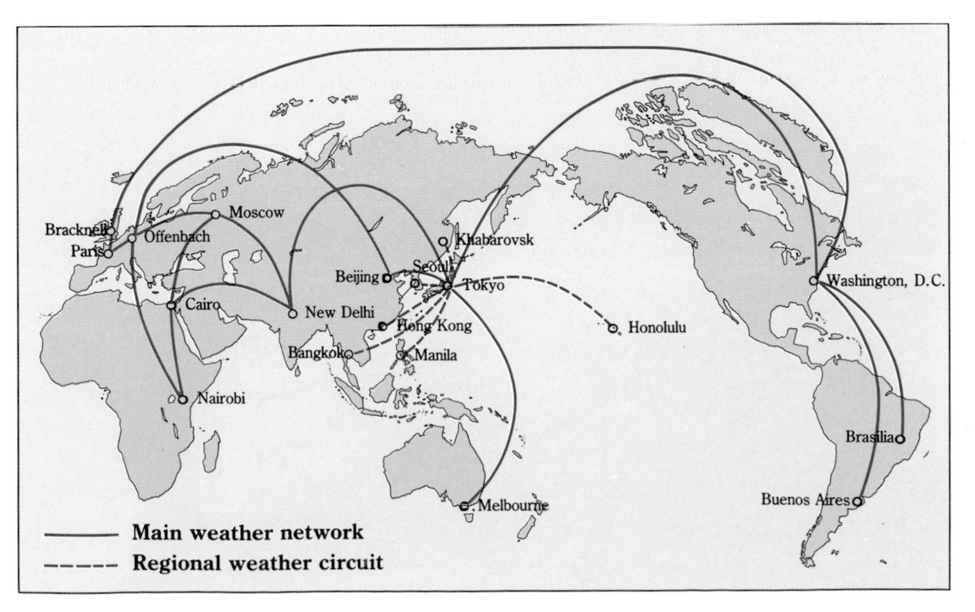

# How Are Weather Maps Made?

Every hour, 10,000 weather reporting stations around the world collect information on local air pressure, temperature, precipitation, wind, and other factors. These facts and figures then travel to national weather centers—in the United States to the National Meteorological Center in Maryland. Meteorologists then add in data from radiosondes, radar, and weather satellites to produce charts of conditions on the surface *(below, center)* and in the upper atmosphere *(right)*.

These current-condition maps are the basic tools that meteorologists use when they forecast the weather. Weather services throughout the world make similar maps; they are able to share them with other countries by means of a global weather network.

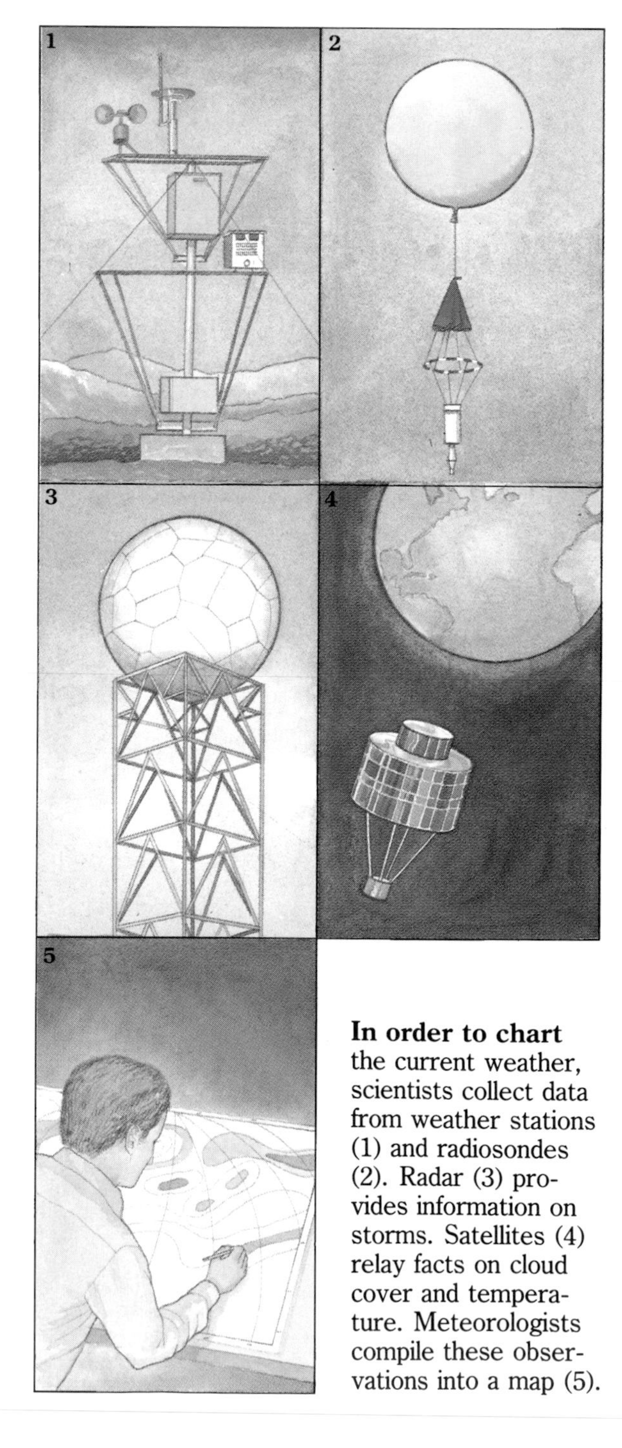


**In order to chart** the current weather, scientists collect data from weather stations (1) and radiosondes (2). Radar (3) provides information on storms. Satellites (4) relay facts on cloud cover and temperature. Meteorologists compile these observations into a map (5).

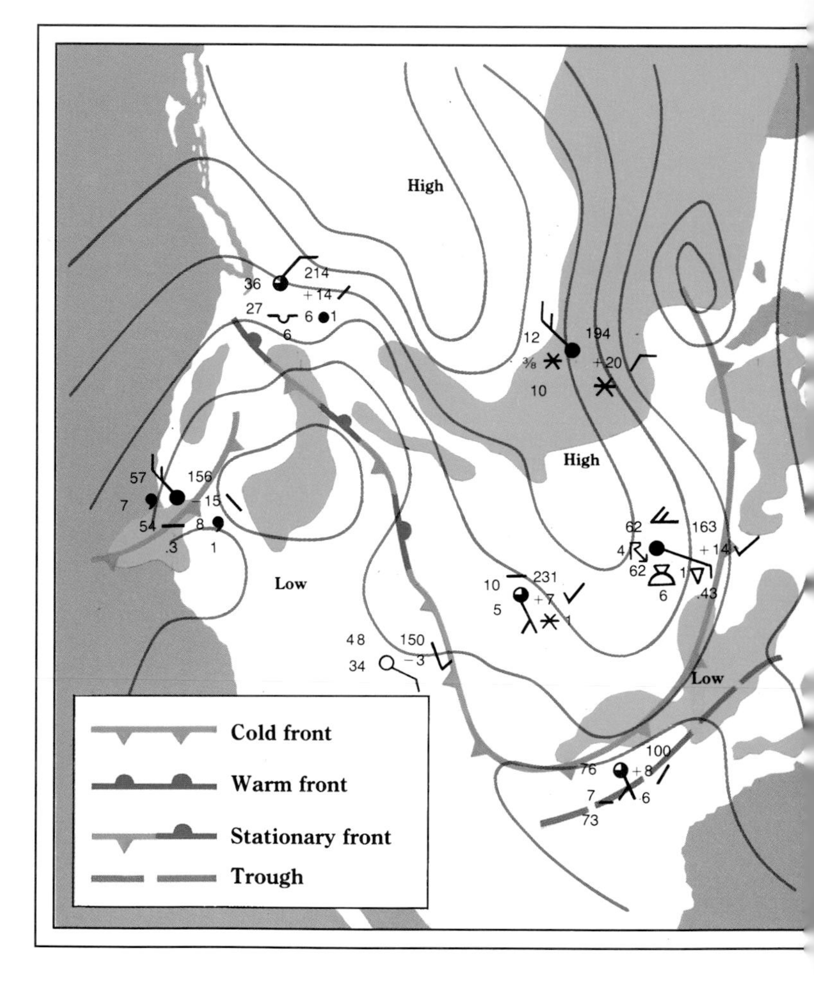


**Surface weather map, October 29**

Mapmakers translate observations from weather stations around the country into station models *(above)*. These models consist of a circle at the location of each station surrounded by symbols showing current and recent weather. (For simplicity, only a few station models are shown here.) Isobars *(black lines)*, fronts *(blue and red lines)*, and areas of precipitation *(green)* are added to complete the map. Produced every hour, these maps allow meteorologists to follow the changing weather.

## A map of the upper air

Because conditions in the upper atmosphere influence weather on the surface, scientists gather information on the upper atmosphere from radiosondes. This enables them to produce a 500-millibar constant-pressure map *(right)* in which black contour lines represent the height at which air pressure equals 500 millibars. Flags show wind speed and direction, and red lines indicate temperature.

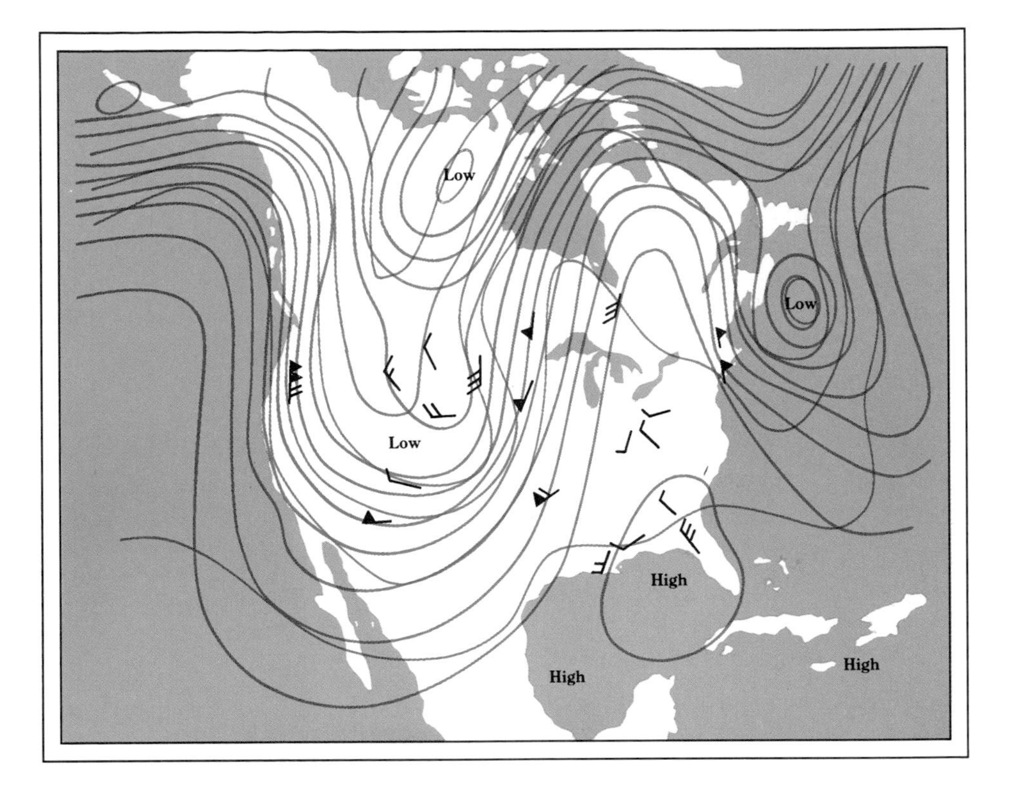


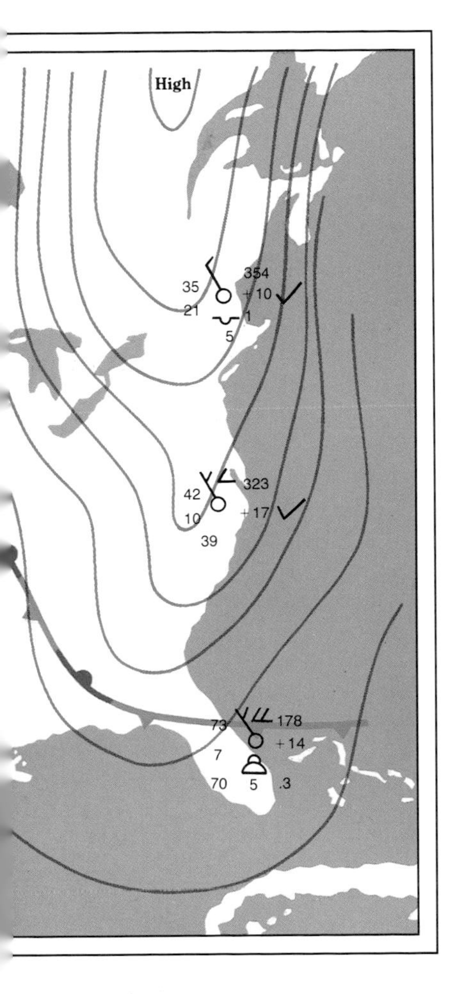


## Station models

A station model *(right)* describes the weather at one location. The right side of the circle summarizes sea-level pressure, dropping the first two numbers; the pressure here is 1014.7 millibars. The left side displays temperature and precipitation. The top shows high- and middle-level cloud types, while the bottom indicates low-level clouds and precipitation in the last three hours.

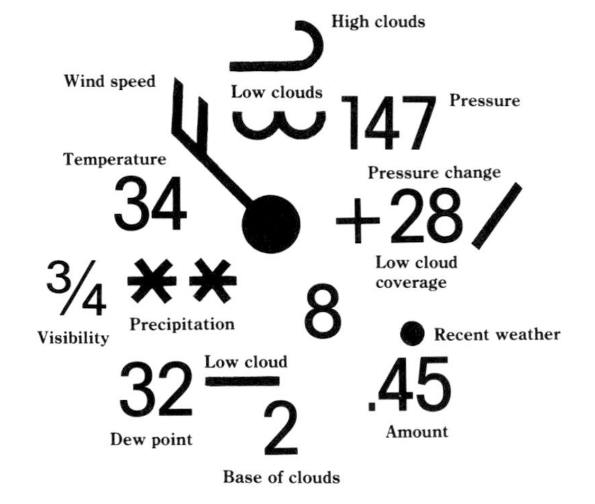


**The weather symbols** at right are used on weather maps. Sky symbols indicate the amount of sky covered by clouds, while current-weather symbols show the type and strength of precipitation. Pressure symbols reflect the trend in pressure since the last reporting period. Clouds are divided into high, middle, and low types. The number of barbs and pennants indicates wind speed; the placement of the flag shows wind direction.

| Sky | Current weather | | Pressure | Cloud type | | Wind |
|---|---|---|---|---|---|---|
| ○ Clear | ● Rain | Freezing rain | — Steady | — Stratus | Altostratus | 5 knots |
| Scattered | Drizzle | Rain showers | \ Falling | Stratocumulus | Altocumulus | 10 knots |
| Half/half | Snow | Ice pellets | / Rising | Nimbostratus | Cirrostratus | 15 knots |
| Broken | Hail | ≡ Fog | Falling, then rising | Cumulus | Cirrocumulus | 20 knots |
| ● Overcast | Thunderstorm | Snow showers | Rising, then falling | Cumulonimbus | Cirrus | 50 knots |

# How Is a Daily Forecast Created?

Once meteorologists know the current weather, they are able to prepare the daily weather forecast. Working out of local weather stations and central weather agencies, the scientists look at surface maps as well as maps of the upper atmosphere, cloud images from weather satellites, radar echo maps, and a variety of other information. After discussion, the meteorologists issue weather forecasts through newspapers, television reports, and other outlets *(right)*.

In addition to the daily forecast, many agencies also come up with precipitation forecasts several times a day. Every day there is also a new prediction for the week beginning on that day. Longer-range forecasts—for one month, three months, and six months—come out regularly. The agencies also produce special reports as needed to warn communities of dangerous weather such as high winds, thunderstorms, tornadoes, heavy snow, and floods.

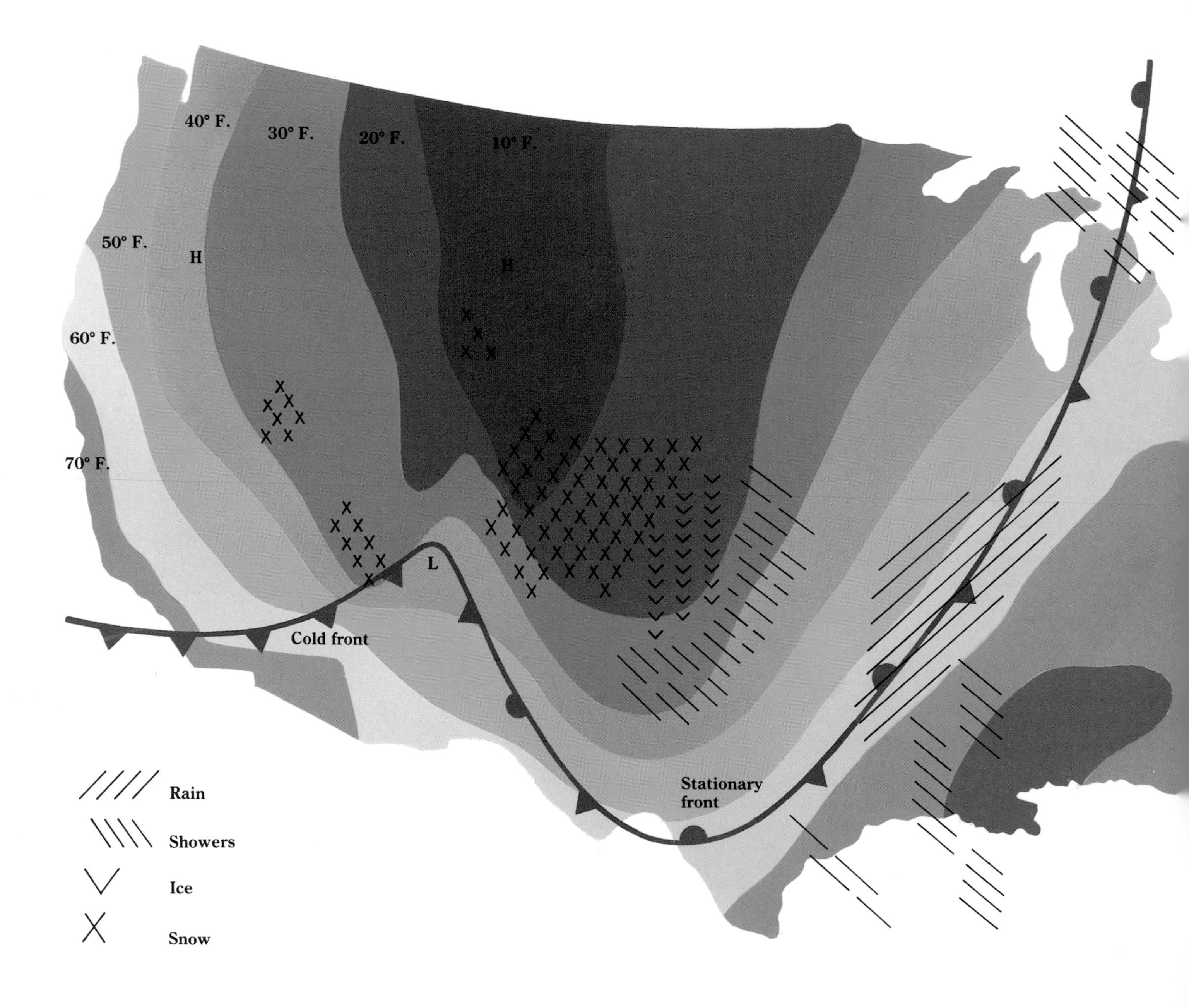

**In a typical year,** 10,000 violent thunderstorms, 5,000 floods, 1,000 tornadoes, and several hurricanes hit the United States. To warn the public, the National Weather Service provides forecasts as well as warnings of severe weather and floods for all of the United States. These forecasts reach the public through newspapers, telephone, radio, and television. The National Weather Service also forecasts the weather for aircraft and shipping operations.

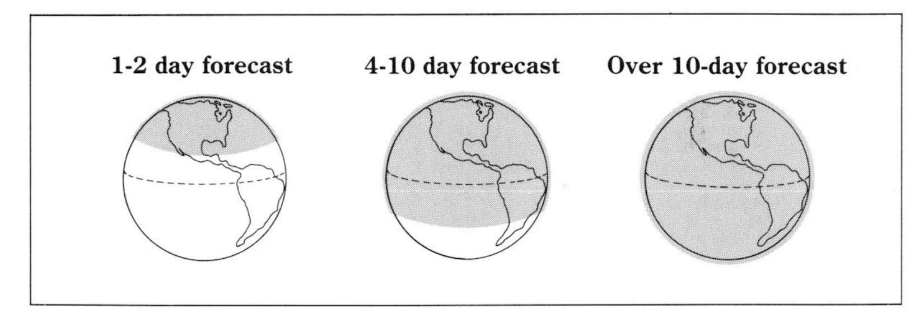


**The longer the range of a forecast,** the greater the amount of information necessary to prepare it *(above)*. For a 1- to 2-day U.S. forecast, scientists need observations of much of North America and surrounding oceans. Four- to 10-day forecasts use data from 20° south latitude to the North Pole. Forecasts past 10 days require information from the entire Earth.

H
40° F.
L
50° F.
60° F.
70° F.
80° F.

**A forecast map** for the day of October 30 *(left)* shows temperatures *(color-coded)*, precipitation, and the location of highs, lows, and fronts. A stationary front bringing rain and cold weather stretches across the country from the Great Lakes to Texas.

**Infrared satellite images** *(right)* depict the progress of a line of storms across the southeastern U.S. White areas show the highest and most dangerous clouds.

# Why Do People Observe Antarctica?

Antarctica, the coldest continent on Earth, plays a major role in determining the world's weather. Nearly all of Antarctica is covered by ice, at an average thickness of 7,000 feet. Low temperatures persist even in sunny weather, since the white surface of the ice reflects most of the energy from the sunlight that strikes it.

The frigid continent acts as a refrigerator for the whole globe, cooling air that flows from lower latitudes. As this air returns to warmer regions, it influences weather far from Antarctica. Thus information on weather at the South Pole is important in the preparation of world weather maps, as well as in research on weather and climate. Several nations maintain a total of 48 observation stations in Antarctica that transmit daily weather observations all over the world.

**Antarctic cloud patterns** highlight regions of low pressure over the continent.

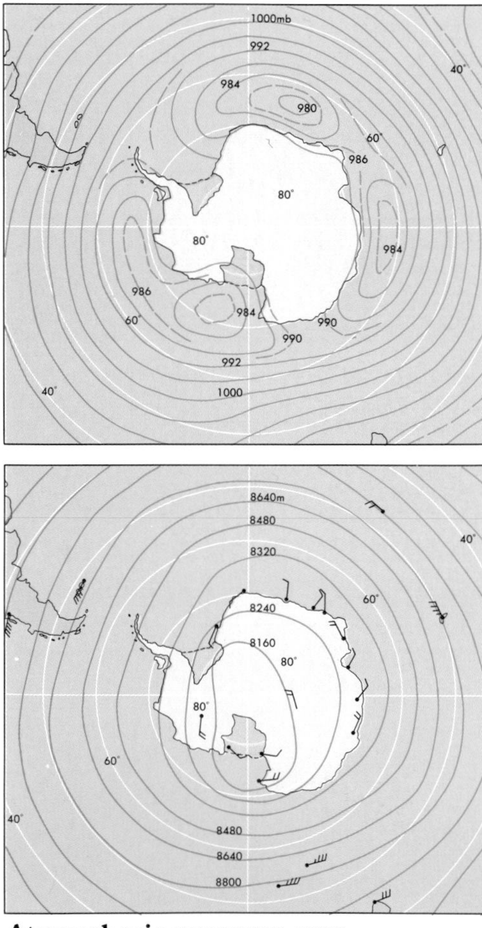


**Atmospheric pressure** over the Antarctic creates air flows that influence weather around the world. The upper map of surface readings shows three areas of low pressure surrounding the continent in July. The lower map shows the different heights (in meters) at which a 300-mb pressure surface can be found.

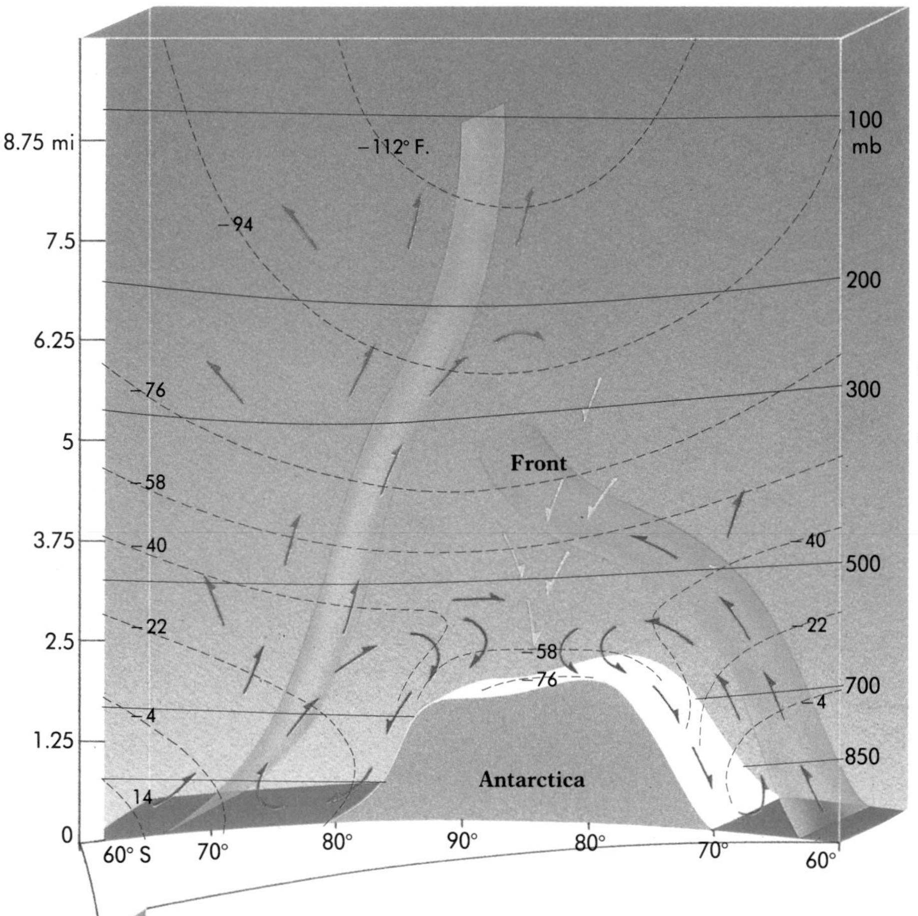


## Antarctic weather

Even in summer, the Antarctic temperature does not rise above freezing; it stays below -20° F. in the interior. Warm air from lower latitudes cools as it is forced upward by the continental landmass; it descends over the central region, then flows down to coastal areas and beyond. The continual recycling of air spreads the cooling power of Antarctica far beyond its shores.

### Reflecting the sun's energy

Dark blue on a world map of albedo (reflective power), Antarctica reflects 80 percent of the solar energy that reaches its surface. The continent absorbs far less solar energy than any other region of similar size.

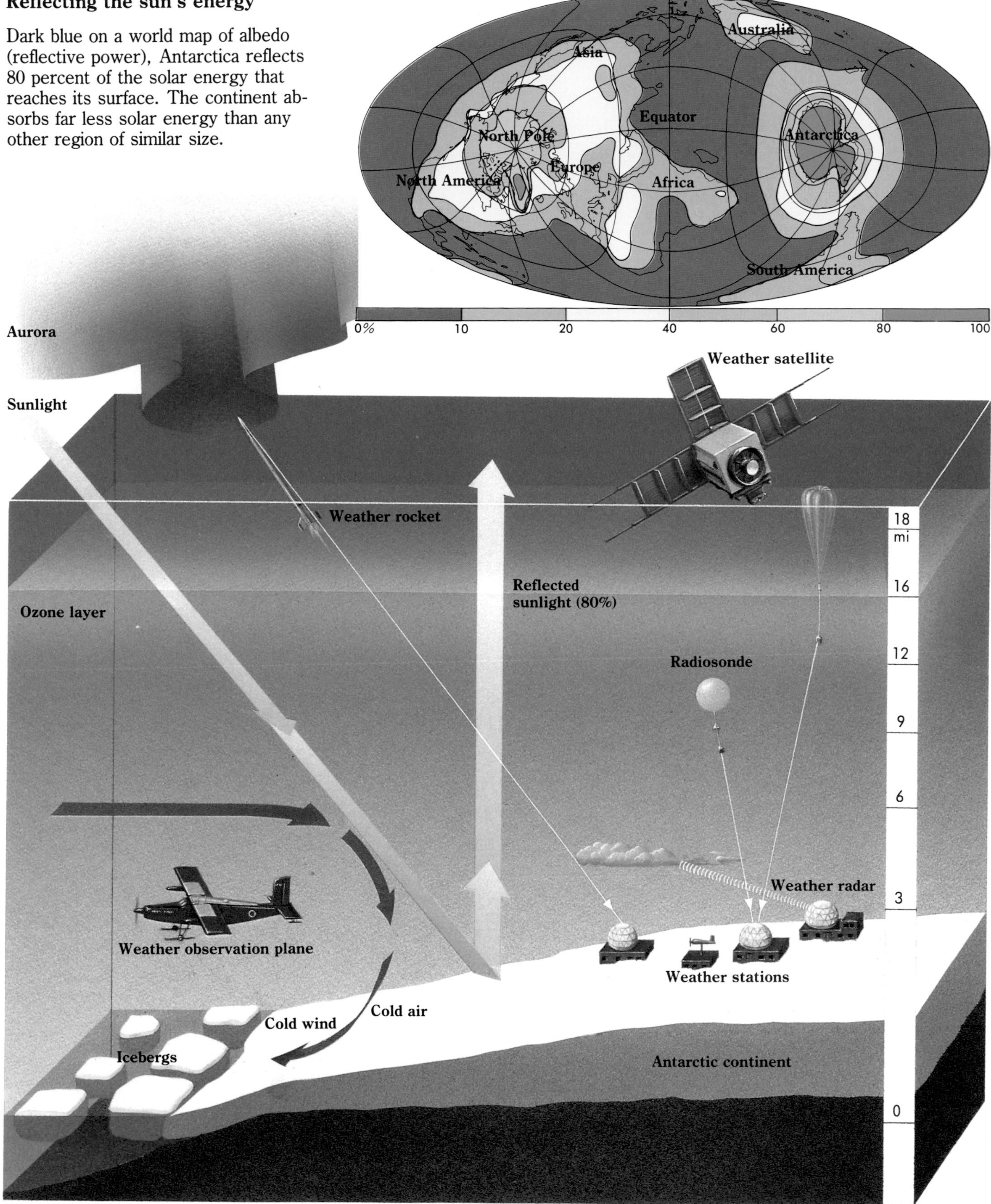


### South Pole weather watch

Weather observation in Antarctica follows the same principles employed in other parts of the world, although the harsh environment often hampers operations. Ground-based observations record temperature, atmospheric pressure, humidity, wind direction and speed, ozone concentration, and amount of sunlight. Balloons, aircraft, and rockets obtain similar information at different levels of the atmosphere and search for suspended particles. Weather radar and satellites track clouds, and satellites also observe the state of snow and ocean ice. Data from these various sources are relayed to meteorological facilities around the world for use in forecasting and research.

# 7

# The Climate of System Earth

A global inequality is responsible for Earth's climate: The sun bathes the tropics in direct light but bestows only oblique rays on the poles. This uneven heating of the planet's surface sets in motion air and water currents that create the world's long-range weather patterns, or climate. Together the currents act like a giant heat engine, moving solar warmth from the equator toward the poles and carrying cooler air and water in the opposite direction.

As the heated equatorial air flows toward the cold polar regions, it is twisted by the Earth's rotation. In the Northern Hemisphere, the air currents are deflected to the right; in the Southern Hemisphere, to the left. This distortion, known as the Coriolis effect, is so consistent that scientists have made maps of worldwide air-circulation patterns in the upper atmosphere and prevailing winds at the Earth's surface. The fastest air currents of all circumnavigate the Earth; called jet streams, they reach top speeds of 200 miles per hour.

Measured for decades and even centuries at points all over the globe, the world's winds, temperatures, and precipitation yield a portrait of a dynamic planetary climate. From the ice ages of the past to the suspected global warming of the present, the one constant in the climate of system Earth has been change itself.

Suggesting the surface of an alien world, wind-sculpted sand dunes stretch toward the horizon of the Sahara. This arid land, which receives less than 4 inches of rainfall a year, represents just one of Earth's many climate extremes.

# Do Ocean Currents Affect Climate?

Covering 72 percent of the planet's surface, oceans are the most commodious heat-storage reservoirs on Earth: The upper 10 feet of the world's oceans hold as much heat as the entire atmosphere. More remarkable still is the ability of ocean currents to transfer that heat. The transatlantic Gulf Stream, for example, moves 4.5 billion cubic feet of sun-warmed seawater per second, transporting more energy northward per hour than would be generated by the burning of 5 billion tons of coal.

This heat transfer has a moderating effect on the climate of lands washed by warm-water currents. Great Britain, the terminus of the Gulf Stream, boasts average winter temperatures 10° to 15° F. higher than Labrador, which has the same latitude but is outside the current's path. And Reykjavik, Iceland, enjoys warmer winters than New York, 2,400 miles to the south.

Not every ocean current, however, improves the climate of lands nearby. The California Current, which channels chilly waters southward from the North Pacific, makes for unusually cool summers along the west coast of the U.S.

**The current-climate connection**

Isotherms—map lines linking sites of equal temperature—snake unevenly across the globe, revealing the dramatic effect of ocean currents on world climate.

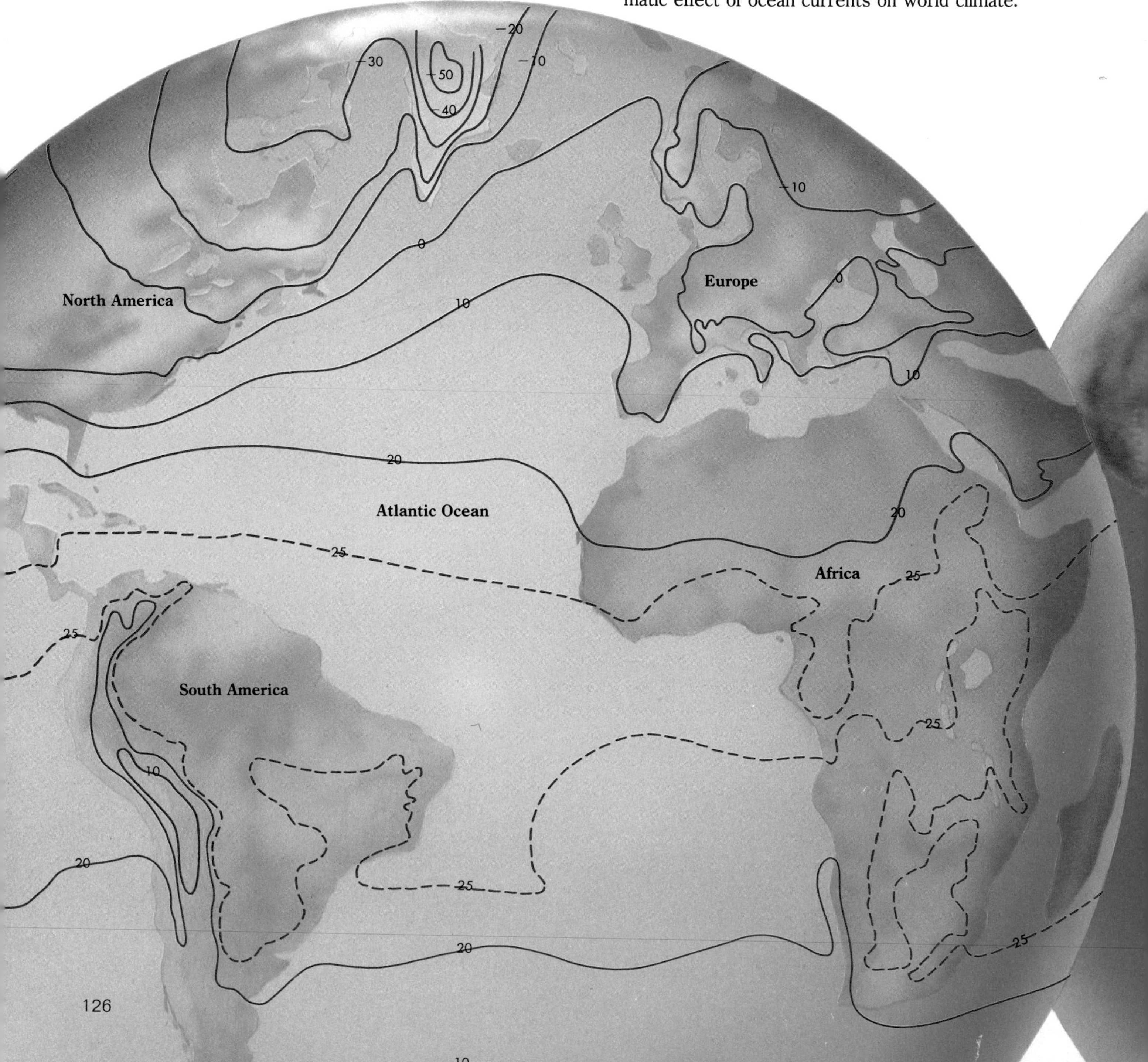

## From America, with warmth

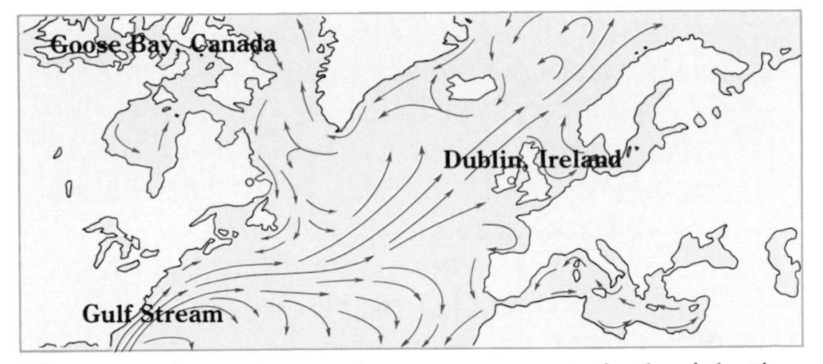


## Currents in the South Pacific



The maps above show major ocean currents in the Atlantic and Pacific oceans. Dublin, warmed by the Gulf Stream, has winter temperatures that are 45° F. higher than Goose Bay. Townsville and Arica have similar temperatures but drastically different precipitation amounts. This is because the cool Humboldt Current chills the air above it, preventing evaporation and thus rainfall, and forming a coastal desert in northern Chile.

## Rivers in the sea

The ocean currents, some of which are shown at left, moderate Earth's climate by carrying warm water away from the equator and cold water away from the poles.

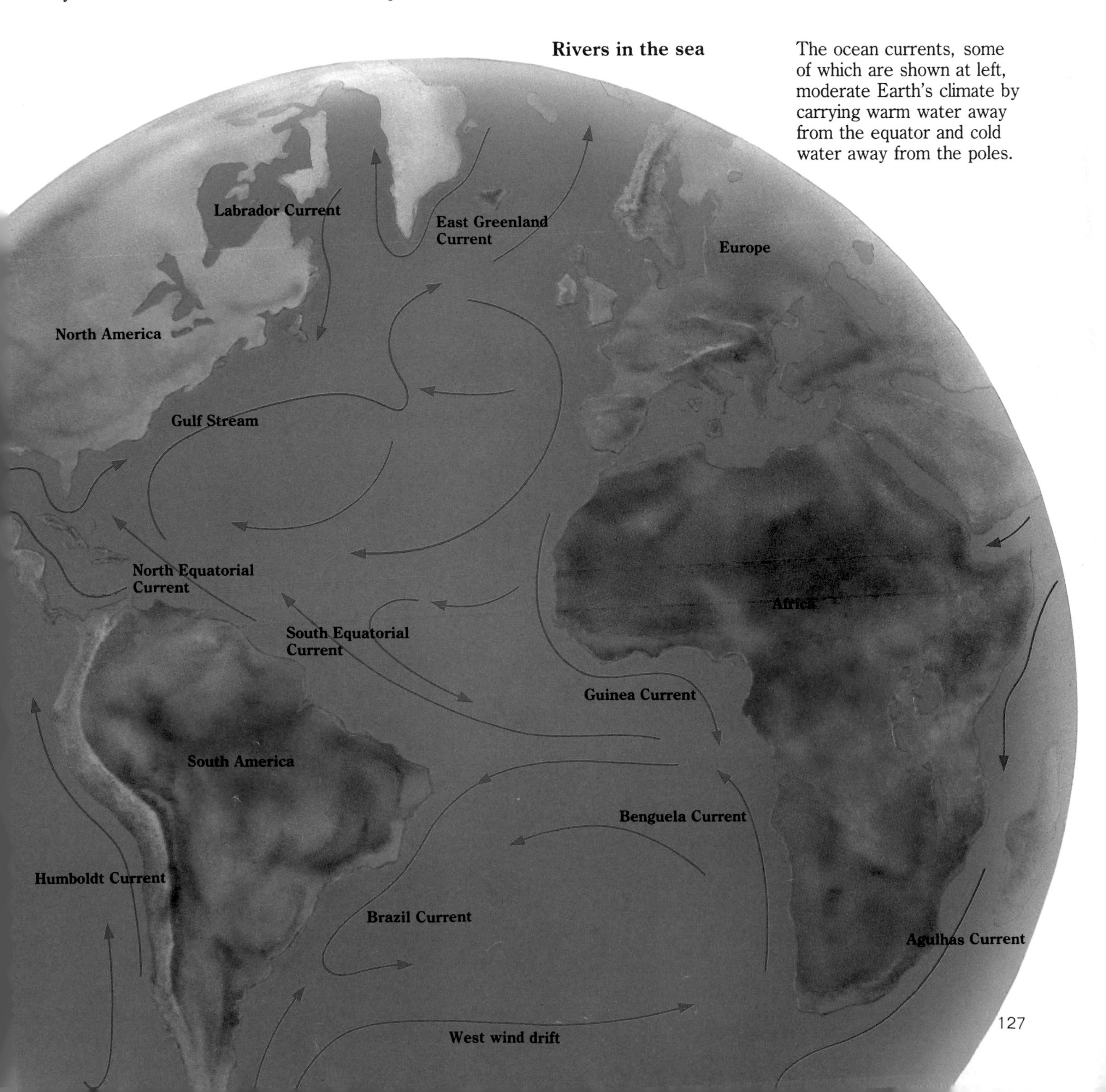

# What Are Climate Zones?

Although no two places on Earth have identical climates, certain climatic similarities make it possible to group widely separated regions into major climate zones. In 1900, German climatologist Wladimir Köppen divided the world's climates into five broad classes: tropical rainy (A), dry (B), warm temperate rainy (C), cold snow forest (D), and polar (E). This classification scheme, illustrated on the map below, has come to be known as the Köppen system.

Köppen based his system on two factors—temperature and precipitation. In his climate

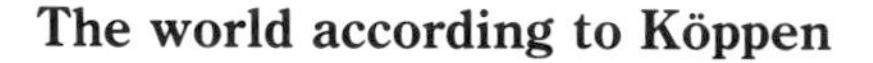

**The world according to Köppen**

One of the oldest climate-mapping schemes, the Köppen classification system, is still in widespread use today. As shown on the color-coded map at right, Köppen suggested that weather patterns the world over belong to one of five major climate zones. Zones A and C have the densest human population. Zone B is the driest, while zone E is the coldest: both have climates so harsh that no trees can grow there.

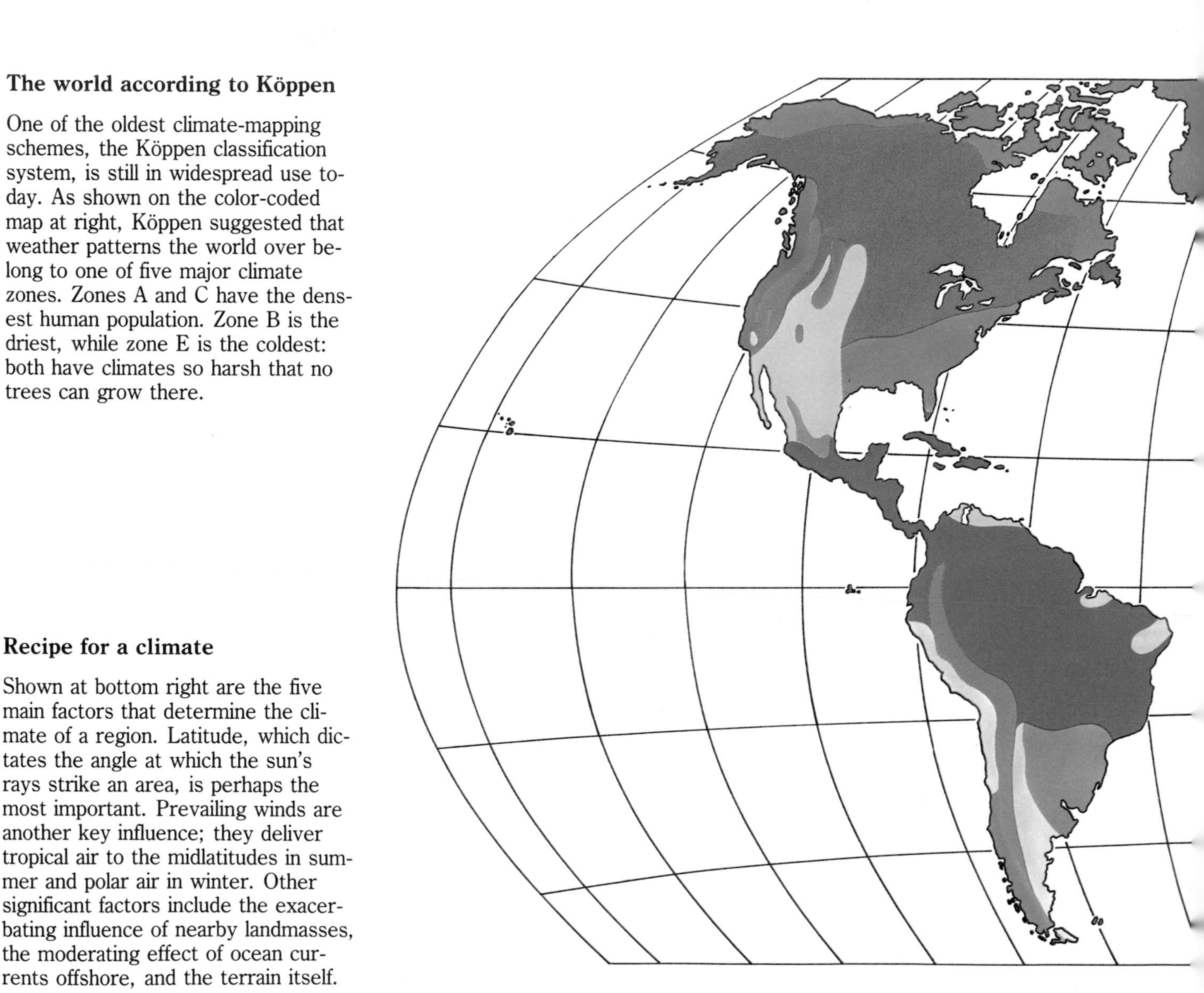


**Recipe for a climate**

Shown at bottom right are the five main factors that determine the climate of a region. Latitude, which dictates the angle at which the sun's rays strike an area, is perhaps the most important. Prevailing winds are another key influence; they deliver tropical air to the midlatitudes in summer and polar air in winter. Other significant factors include the exacerbating influence of nearby landmasses, the moderating effect of ocean currents offshore, and the terrain itself.

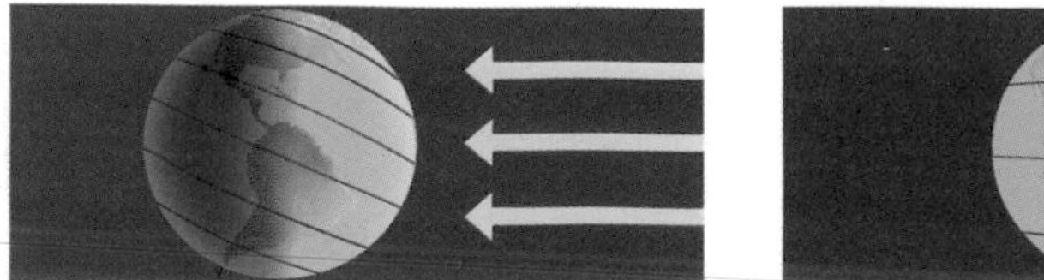

**Latitude**

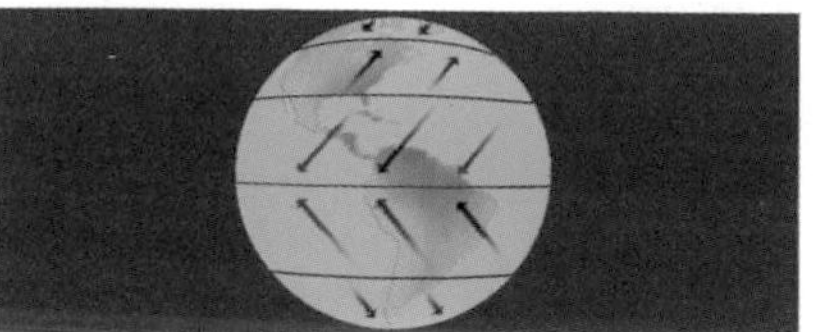

**Prevailing winds**

zone A, which includes the rain forest and the savanna, the coldest month has an average daily temperature higher than 64° F. In climate zone B, evaporation exceeds precipitation, producing deserts or steppes. Zone C, which includes the eastern United States and western Europe, has an average daily temperature between 26° and 64° F. during its coldest month. Unlike the other groups, zone D climates occur only in the Northern Hemisphere, where they cover landmasses at high latitudes. Finally, climates in zone E have temperatures that never rise above 50° F.

**Warm temperate rainy (C)** **Cold snow forest (D)** **Polar (E)**

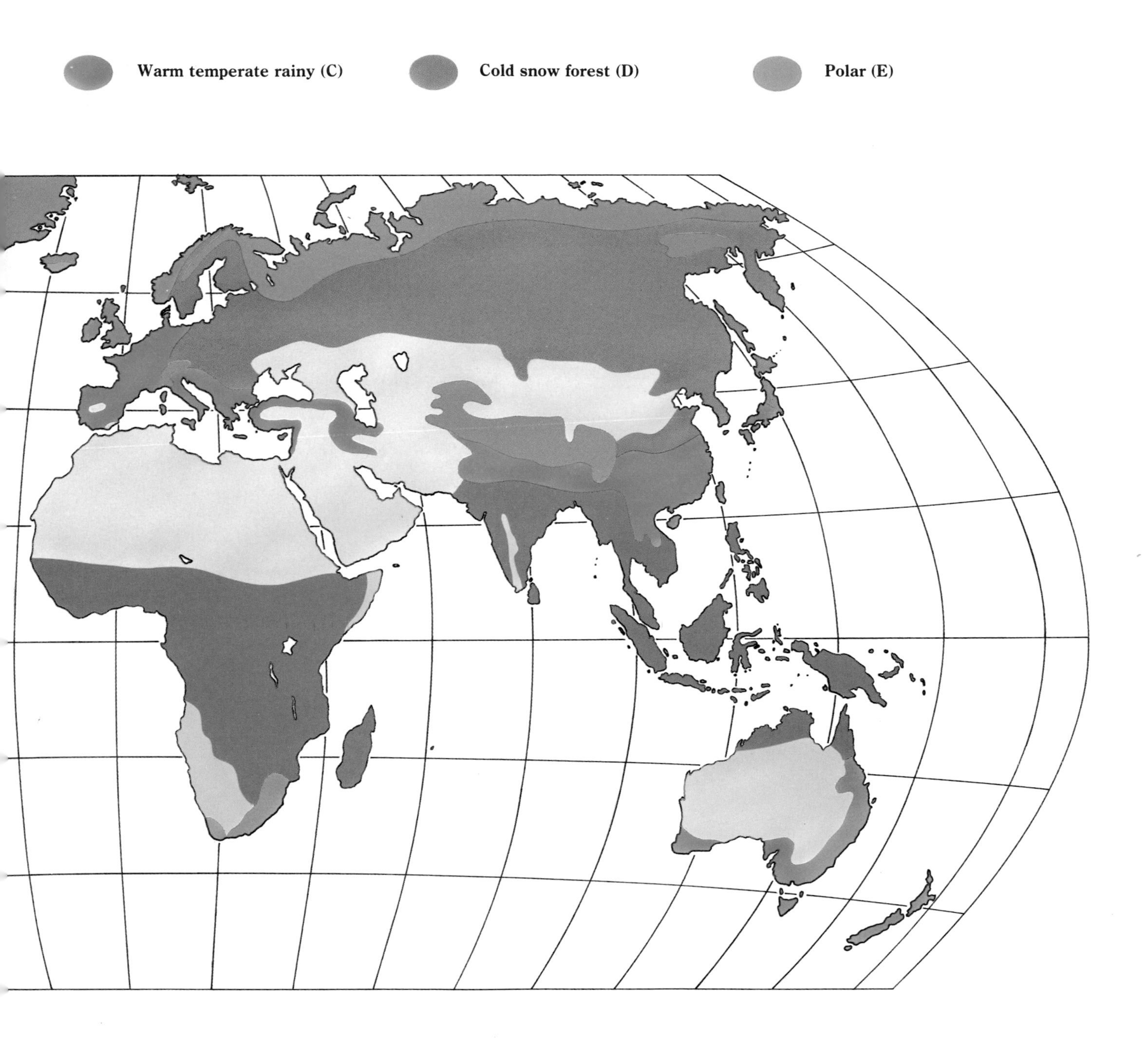

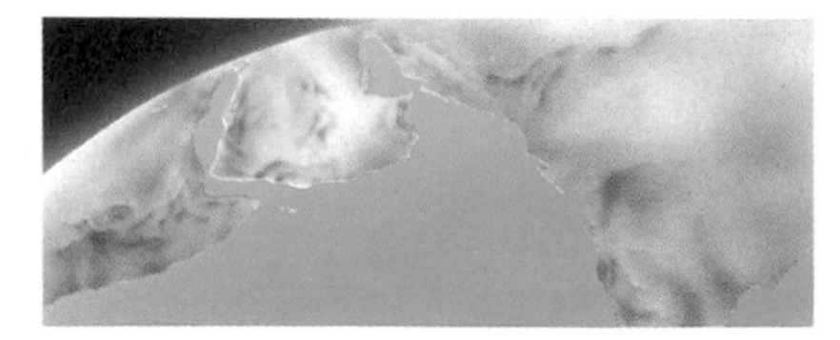

**Distribution of continents**

**Effect of ocean currents**

**Topography**

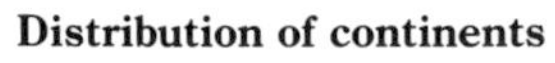

# What Is a Monsoon?

A monsoon is a wind that changes directions with the seasons, blowing moist air from the ocean to the land in summer and blowing dry air offshore in winter. Monsoons occur all across Africa and Asia, but nowhere do they play a more dramatic role than on the subcontinent of India; there, between the months of June and September, the summer monsoon brings steady rains that nourish rice and other crops on which one-tenth of the world's population survives.

Because land heats up faster than water, by May the Asian landmass is often 18° F. warmer than the Indian Ocean. The heated air above the continent rises and expands, creating an area of low air pressure that sets the monsoon in motion. To equalize the pressure, the colder, denser air above the ocean begins flowing inland. Along the way, it picks up evaporating seawater. When this moisture-laden monsoon reaches land, warmer air makes the water vapor condense, and the rains come pouring down.

In winter, the pattern is reversed. The land cools off faster than the ocean, so the monsoon rushes from the interior of the continent toward the sea. For India, the result is sustained dry weather that lasts from October to May.

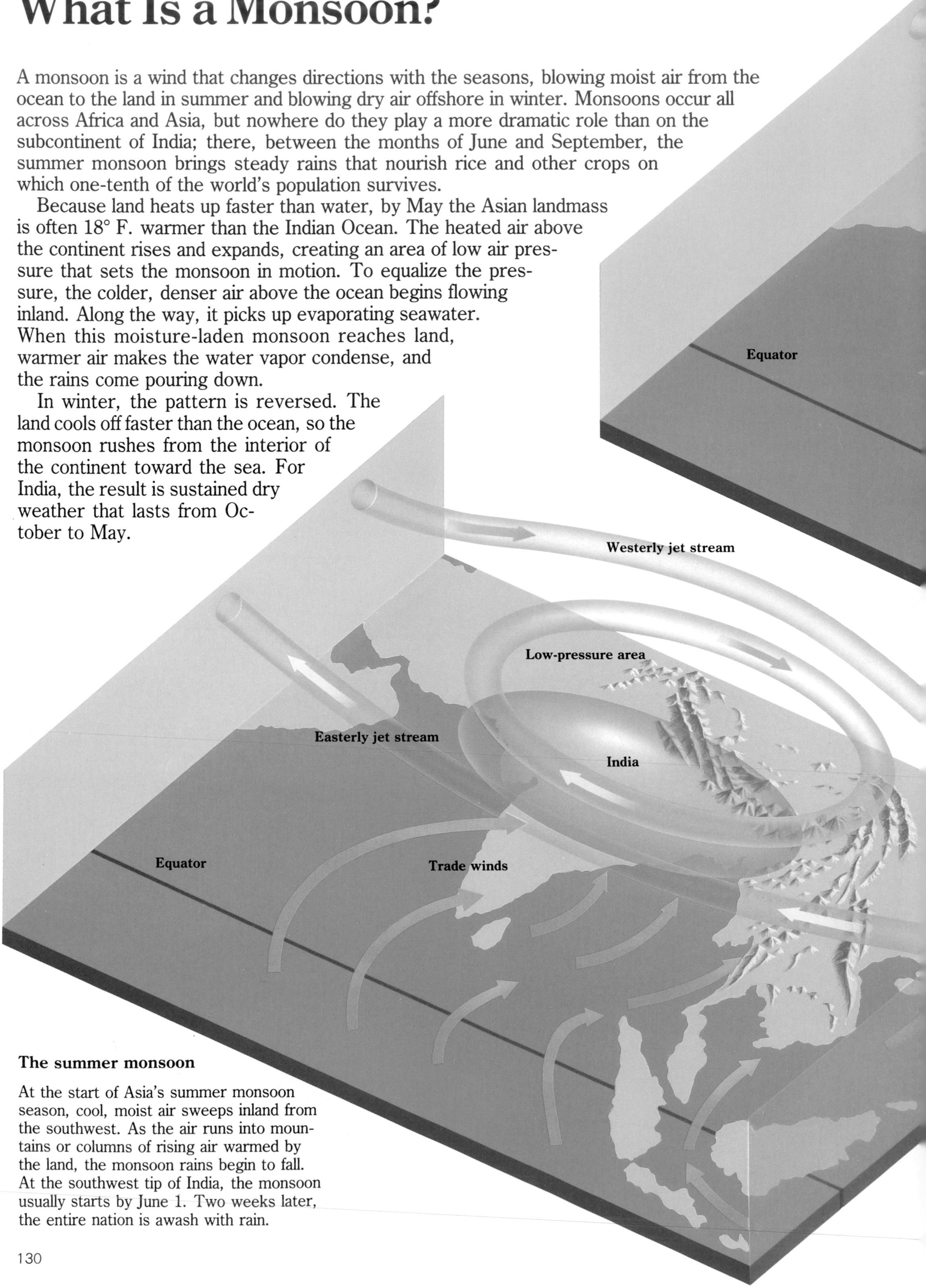

Equator

Westerly jet stream

Low-pressure area

Easterly jet stream

India

Equator

Trade winds

**The summer monsoon**

At the start of Asia's summer monsoon season, cool, moist air sweeps inland from the southwest. As the air runs into mountains or columns of rising air warmed by the land, the monsoon rains begin to fall. At the southwest tip of India, the monsoon usually starts by June 1. Two weeks later, the entire nation is awash with rain.

Westerly jet stream

Westerly jet stream

High-pressure area

Himalayas

Tibetan plateau

China

### The winter monsoon

In winter, the monsoon blows frigid Siberian air toward the coast and out to sea. As this cold air meets warmer air above the oceans, heavy rainstorms drench coastal areas. For this reason, most of Southeast Asia, northern Australia, and the northeastern coastlines of Indonesia get ample rainfall in winter.

### It's a wet, wet, wet, wet world

Monsoons usher in some of the world's wettest weather. They bring the heaviest and most regular rains to East Africa, India, and Southeast Asia.

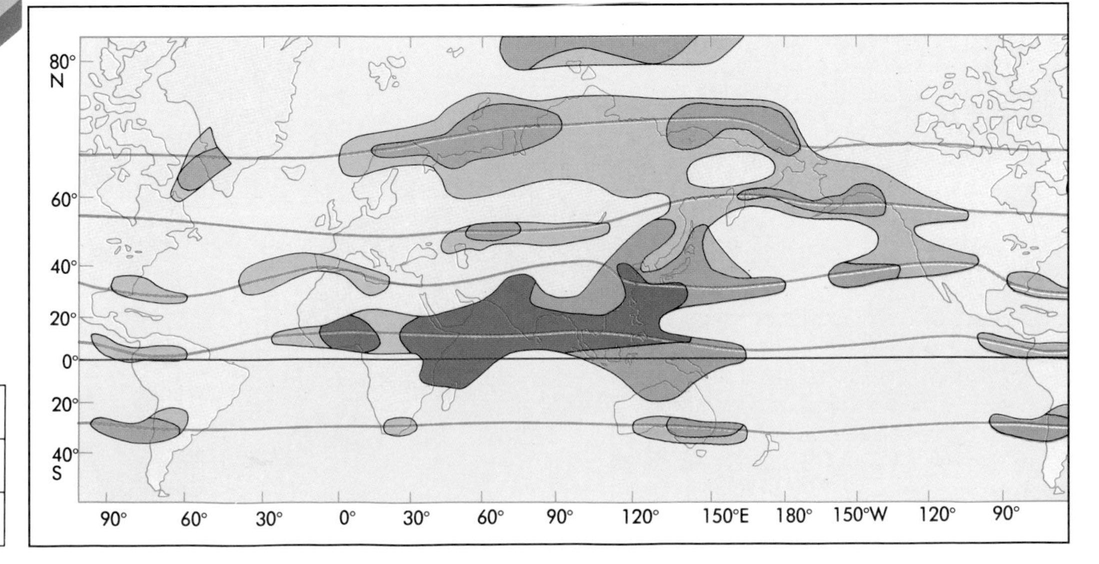


**Annual precipitation**

| | |
|---|---|
| | Very heavy |
| | Heavy |
| | Light |

# Why Does Asia Have Rainy Seasons?

The climate of Asia is distinguished by the seasonal collision of certain prevailing winds. These atmospheric encounters cause the Philippines, Southeast Asia, northern India, southern China, Japan, and the southern tip of Korea to experience a rainy season.

When trade winds blowing northwest toward the equator meet trade winds blowing southwest, the two wind systems form a line of ascending currents known as the equatorial convergence zone. The rising air cools and condenses, then clouds build up and the rains pour down. Yet the convergence zone is hardly stationary; it migrates back and forth across the equator, carrying rainy weather wherever it goes.

Though the convergence zone never travels more than 25° north or south of the equator, areas beyond its reach can still expect a regular drenching. Japan's rainy season, for example, is brought about by the clash of cold and warm air masses in the ocean nearby.

**Two Thai villagers** cross a rain-swollen river during the rainy season. Many homes are built on stilts to keep them above the flood line.

**A migrating cloud bank**

In May, the equatorial convergence zone—where trade winds from the southeast meet those from the northeast—shifts northward, covering the southern tips of Asia. As shown below, thick cloud masses called equatorial waves develop, and the rainy season begins.

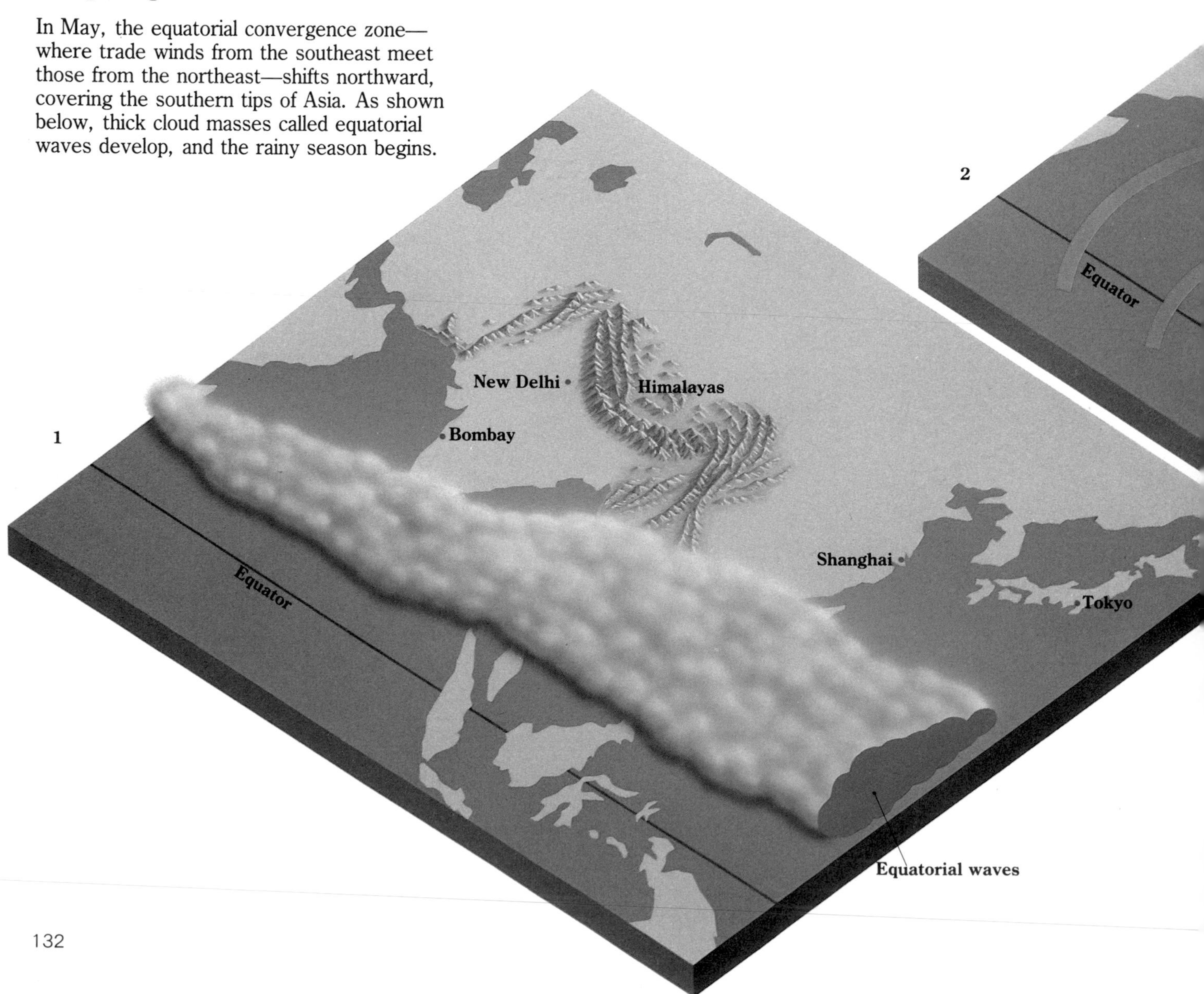

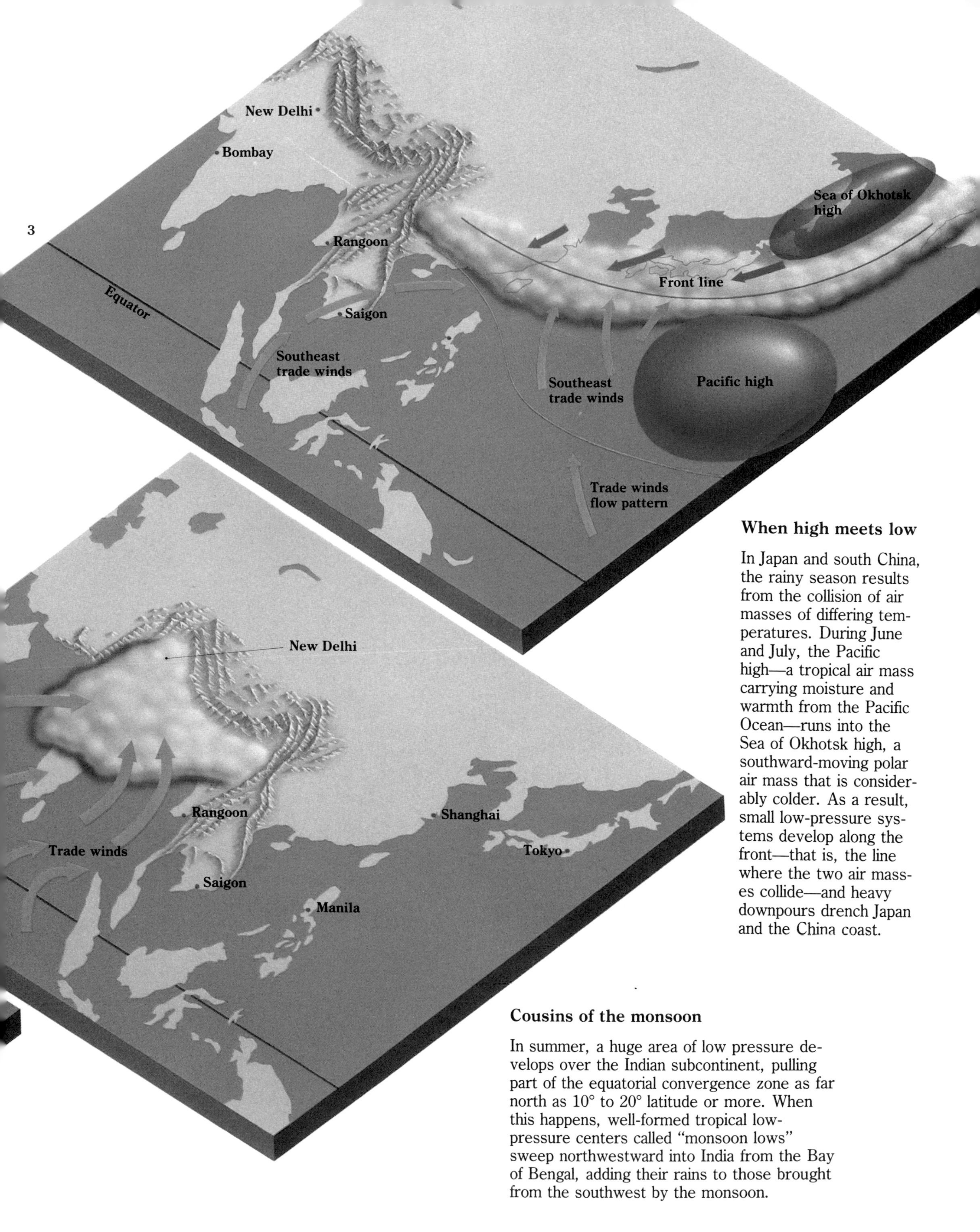


### When high meets low

In Japan and south China, the rainy season results from the collision of air masses of differing temperatures. During June and July, the Pacific high—a tropical air mass carrying moisture and warmth from the Pacific Ocean—runs into the Sea of Okhotsk high, a southward-moving polar air mass that is considerably colder. As a result, small low-pressure systems develop along the front—that is, the line where the two air masses collide—and heavy downpours drench Japan and the China coast.

### Cousins of the monsoon

In summer, a huge area of low pressure develops over the Indian subcontinent, pulling part of the equatorial convergence zone as far north as 10° to 20° latitude or more. When this happens, well-formed tropical low-pressure centers called "monsoon lows" sweep northwestward into India from the Bay of Bengal, adding their rains to those brought from the southwest by the monsoon.

# Why Are Deserts So Dry?

All deserts are dry—they receive less than 10 inches of rain a year—but the reasons for this aridity differ greatly. Subtropical deserts—those between the latitudes of 15° and 35° on either side of the equator—occur in areas where high-pressure air masses are constantly flowing down on the Earth's surface. As these columns of heavy air move toward the ground, they become compressed. The compression warms them up, so the air absorbs moisture rather than releasing it as rain. Mountains, too, can cause deserts. When a soggy air mass hits the wall of a mountain chain, it is forced upward; the air cools and the moisture condenses into rain. Robbed of this precipitation, the leeward side develops into a desert. Even an ocean can create a desert. Wet winds that pass over cold sea currents lose their moisture as rain at sea, depriving the land nearby.

### Hearts of dryness

In temperate climates, land on the windward side of mountains gets plentiful precipitation. Prevailing winds drive moisture-laden air up the mountainsides, causing the air to dump its cargo of rain or snow. The dry air then moves on, shedding little rain on the lands beyond. The result is rain-shadow deserts like the Gobi in central Asia.

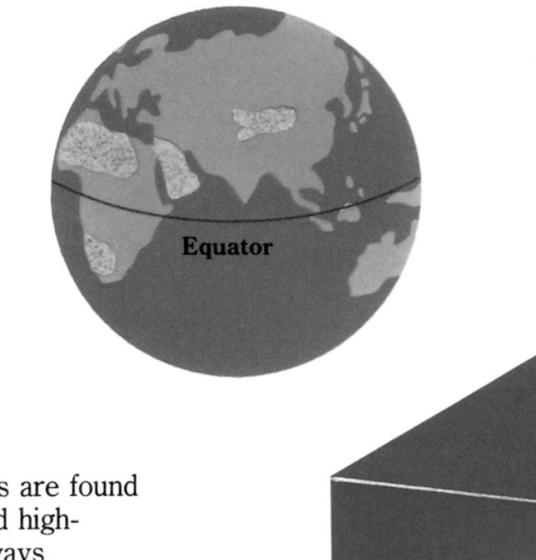


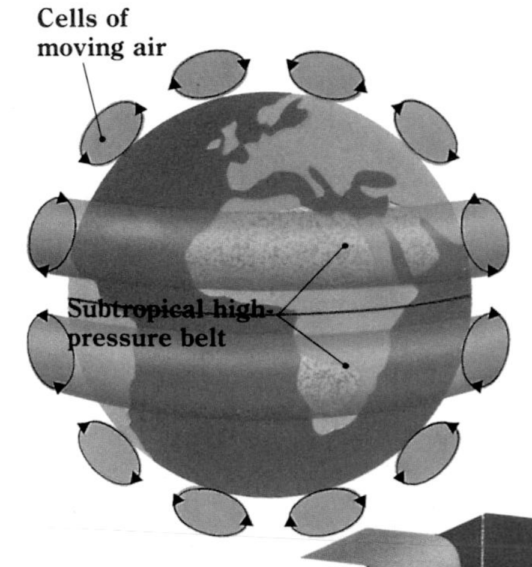


### Dry belts on a wet world

The deserts of the subtropics are found in regions where parched and high-pressure air currents are always descending. (Ascending air currents, by contrast, produce rain.) Two high-pressure belts, each located between about 15° and 35° latitude, bracket the equator *(left)*. The world's largest deserts—the Sahara, the Arabian, and the Australian—lie within these belts.

## Deserts by the sea

Where a constant wind blows along the west coast of a continent, it pushes the surface ocean current out to sea and draws cold water upward from the ocean floor. Any air mass crossing this stretch of frigid seawater becomes chilled and gives up its moisture offshore. A frequent by-product of this interaction is dense fog that hovers above coastal deserts.

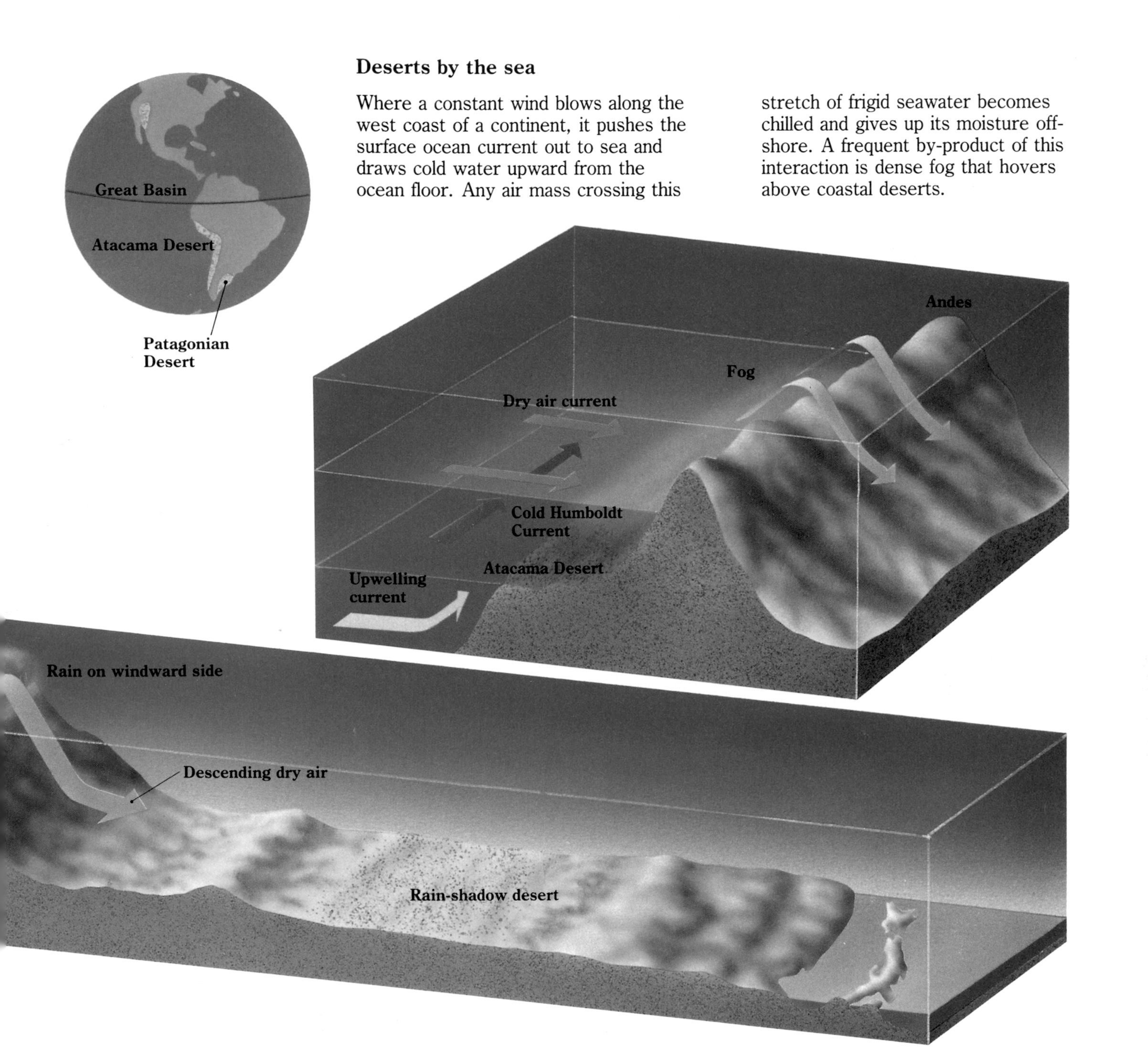

**A camel caravan** crosses the Sahara, which is littered with rocks and pebbles. Sand dunes cover only 12 percent of arid lands.

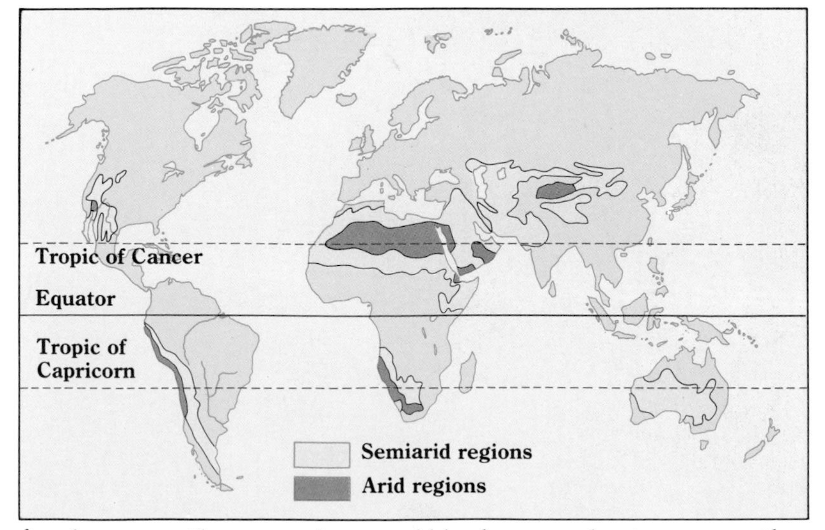


**As shown on the map above,** arid lands cover about one-seventh of the world's solid surface. The majority of deserts are located between 10° and 50° latitude, where they are most likely to form on the western edge of a continent.

# What Is the Greenhouse Effect?

Since the mid-nineteenth century, humanity has become increasingly dependent on the burning of fossil fuels—coal, oil, and gas—to provide energy. In the 1970s, however, scientists discovered that the use of these fuels may be disastrous for the future of the Earth.

The problem is carbon dioxide, a gas emitted when fossil fuels are burned. In the atmosphere, carbon dioxide allows solar energy to strike the Earth but keeps that energy from radiating back into space. This results in a heat buildup that is commonly known as the greenhouse effect.

Many scientists believe that increased levels of carbon dioxide will cause global warming—a significant rise in Earth's average temperature. Over the next few hundred years, global warming could elevate Earth's temperature by a few degrees. This would be enough to partly melt the polar icecaps, which in turn would raise the sea level and flood many of the world's coastal cities. To reduce this possibility, many countries are making their cars more fuel efficient and taking other steps to curtail the amount of greenhouse gases released into the atmosphere.

**The greenhouse effect at work**

Incoming solar energy *(yellow arrows)* has short wavelengths and passes easily through Earth's atmosphere. As this energy warms the Earth, the planet radiates much of it back as long-wavelength infrared energy *(orange arrows)*. But infrared energy cannot pass through carbon dioxide; much of it is trapped in the atmosphere. The result: an increase in planetary temperatures.

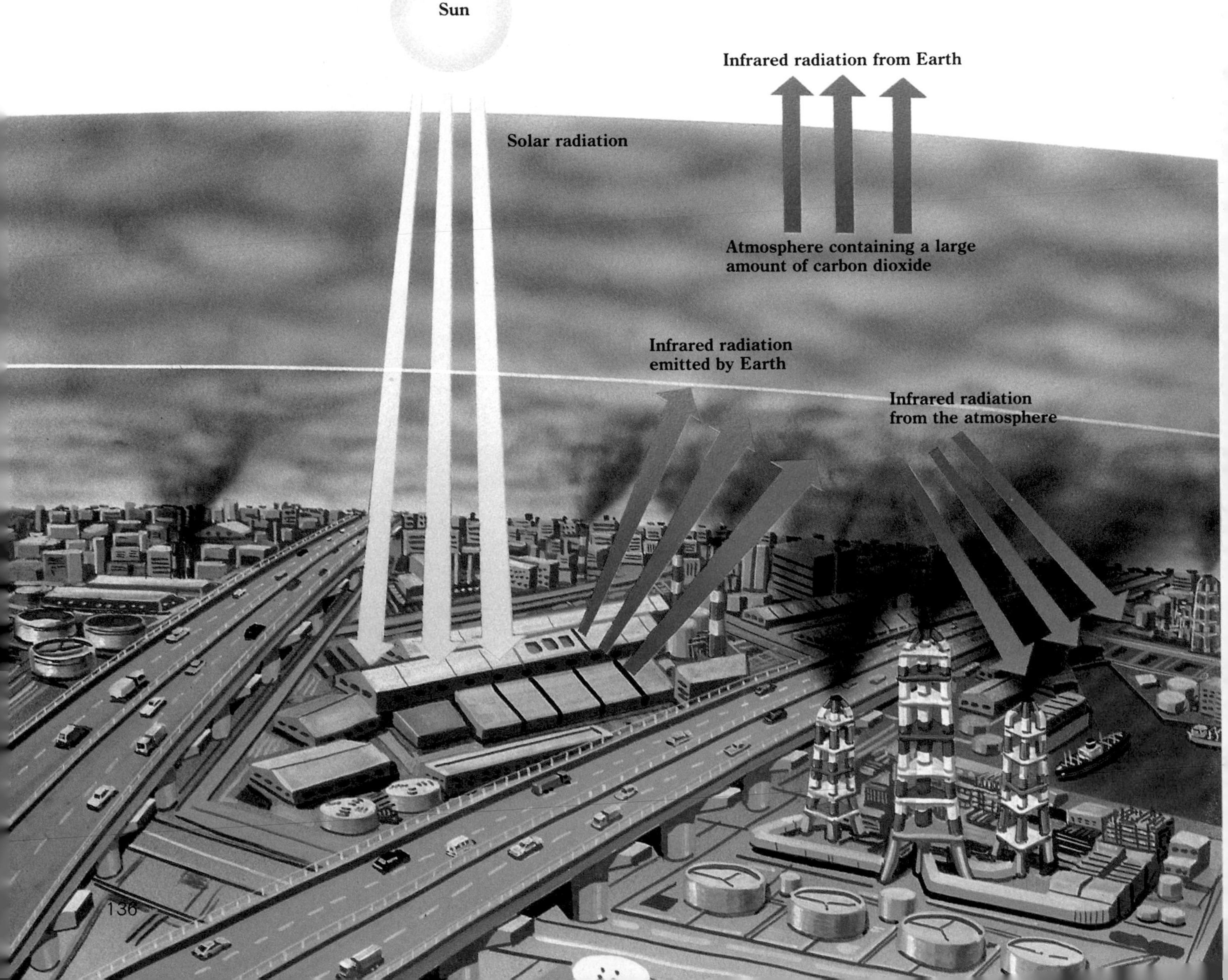

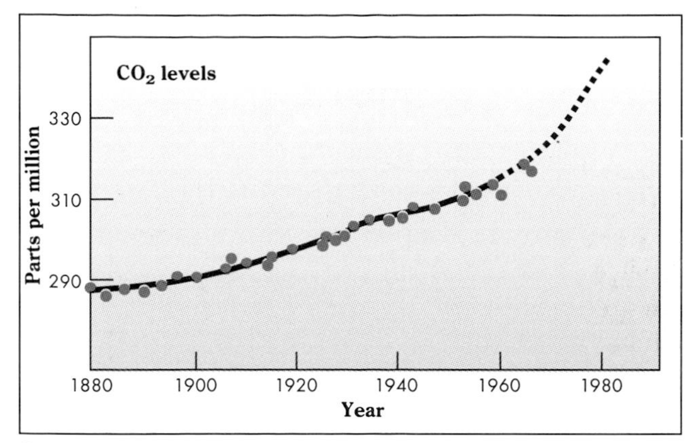


**Changes in carbon dioxide concentration**

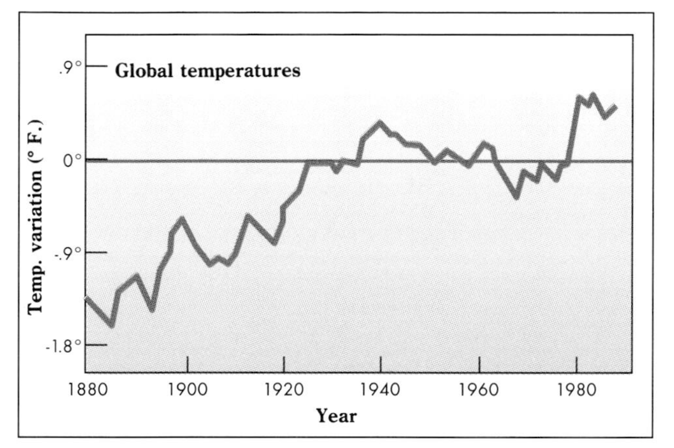


**Global temperature variations**

### The $CO_2$ connection

By analyzing air bubbles trapped in Antarctic ice, scientists can measure how the carbon dioxide content of the atmosphere has changed over the decades. The results, shown in the top graph at left, reveal that concentrations of the gas have skyrocketed. The bottom graph of the Earth's temperature over the same timespan shows a corresponding rise in atmospheric temperatures of 1.8° F. since 1880.

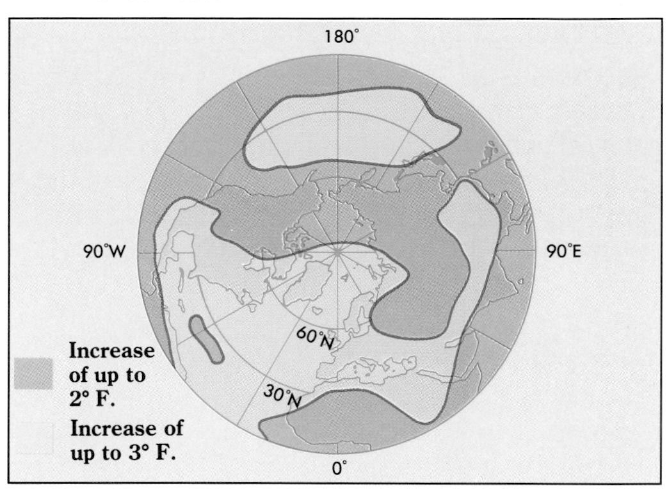


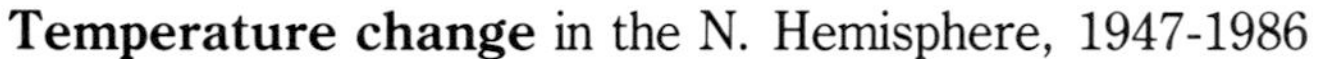

**Temperature change** in the N. Hemisphere, 1947-1986

### Without the greenhouse effect

Although the greenhouse effect will pose dangers if it is allowed to develop unchecked, it is in fact essential for keeping the Earth warm. Without carbon dioxide in the atmosphere, all of Earth's infrared energy would be radiated into space, leaving the planet too cold to sustain life.

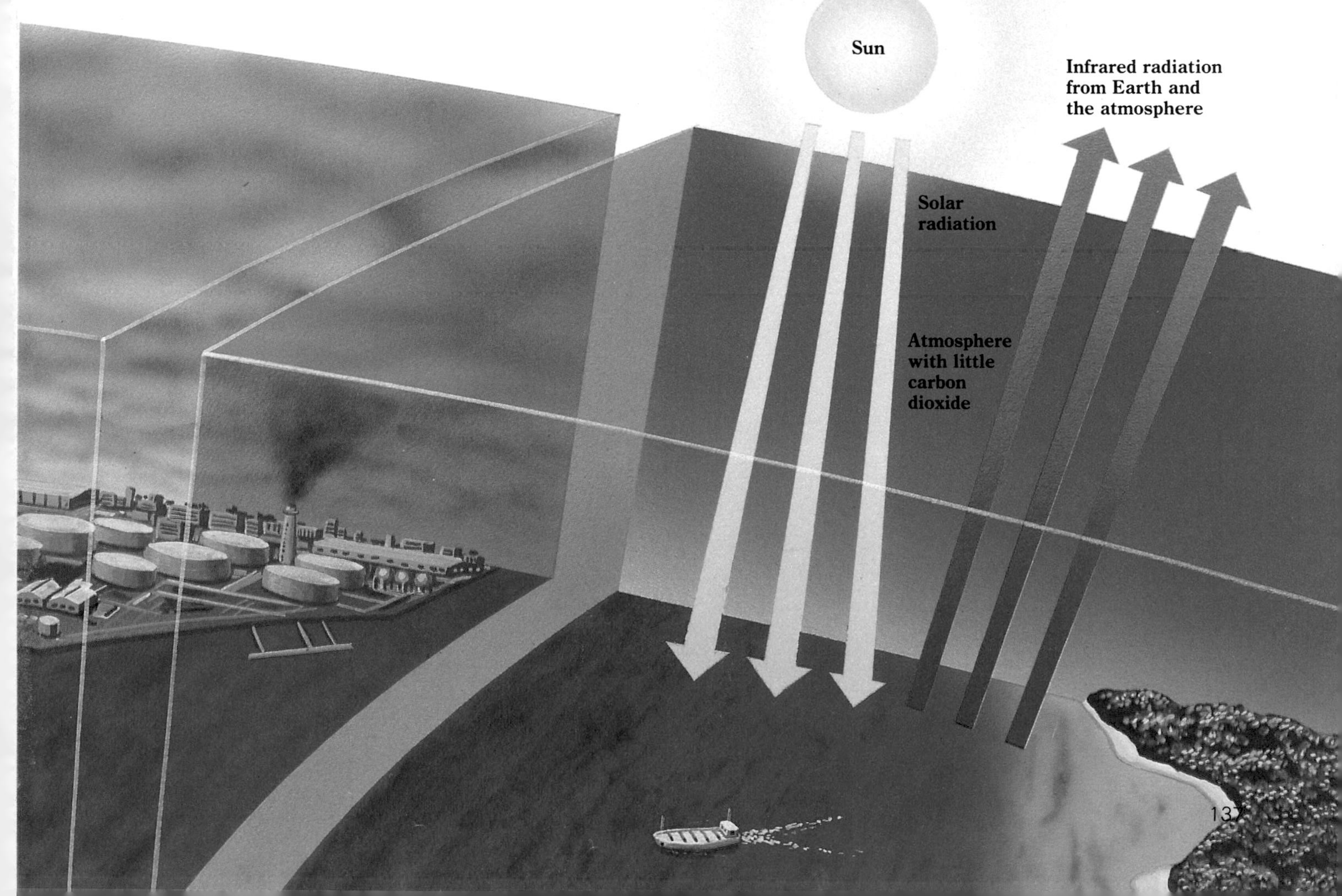

# Why Is a City Warmer Than a Suburb?

With its densely packed skyscrapers and vast expanses of asphalt and concrete, a city interacts with the atmosphere differently than a suburb. The city disrupts air circulation patterns and hinders the evaporation of rainwater. Through auto emissions, the heating of buildings, and the operation of factories, it also throws off a tremendous amount of heat. As a result, a heat island, or inversion, tends to surround the city. The heat island is usually about 10° F. warmer than the air over nearby suburbs; between a large city and its environs, that difference may exceed 15°.

As cities grow ever larger, the heat-island effect could drastically alter the Earth's atmospheric patterns. Some scientists contend that the heat islands generated by growing urbanization combine with the greenhouse effect to further global warming.

### A hot time in the old town

The heat-island effect creates a dome of warm air around a city. High above the city, this hot air blows out toward the suburbs; lower down, the cooler air from the suburbs flows in toward the city's center. Once the cooler air is inside the city it begins to heat up, perpetuating the heat island.

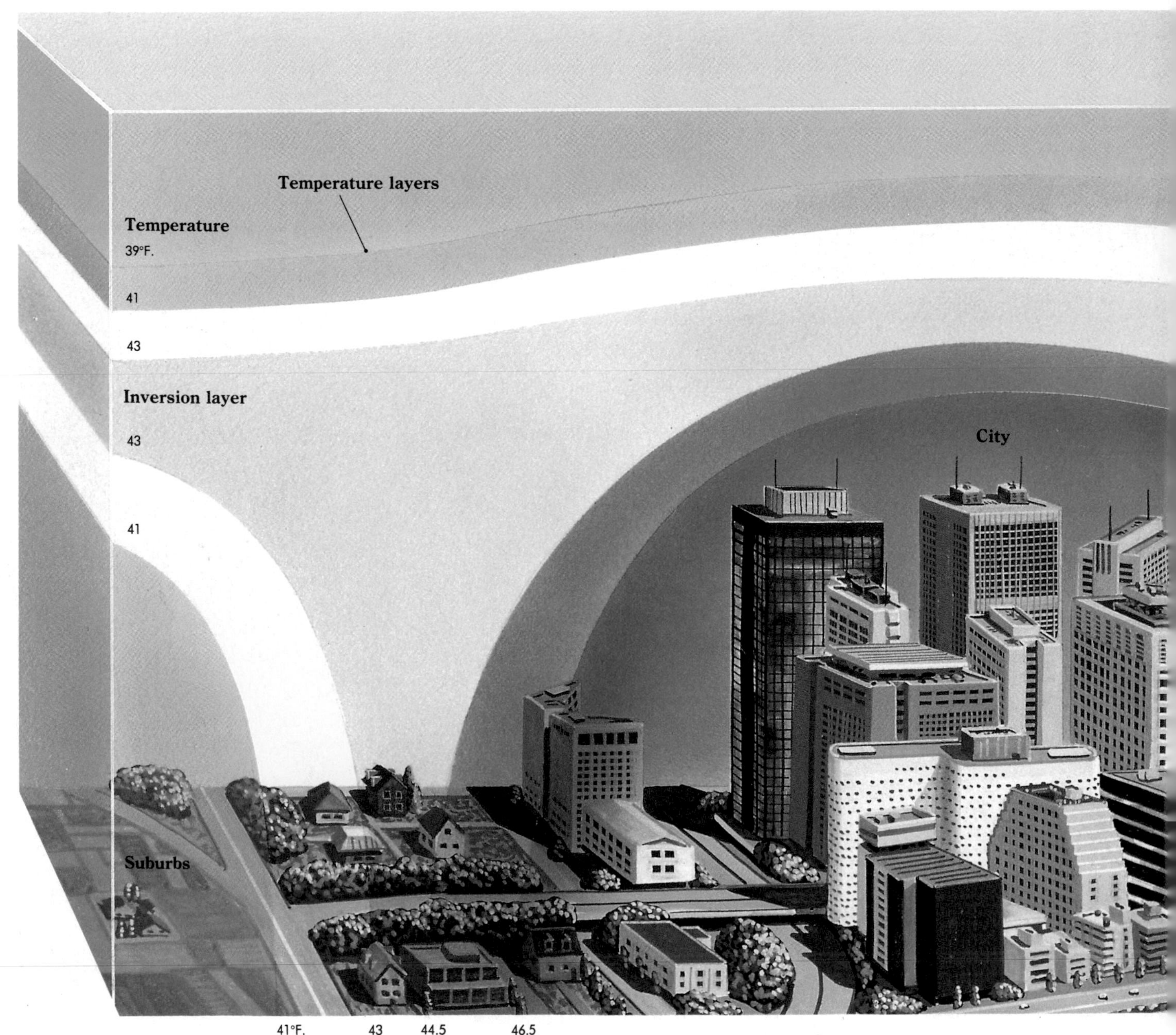

## Tokyo's inversion

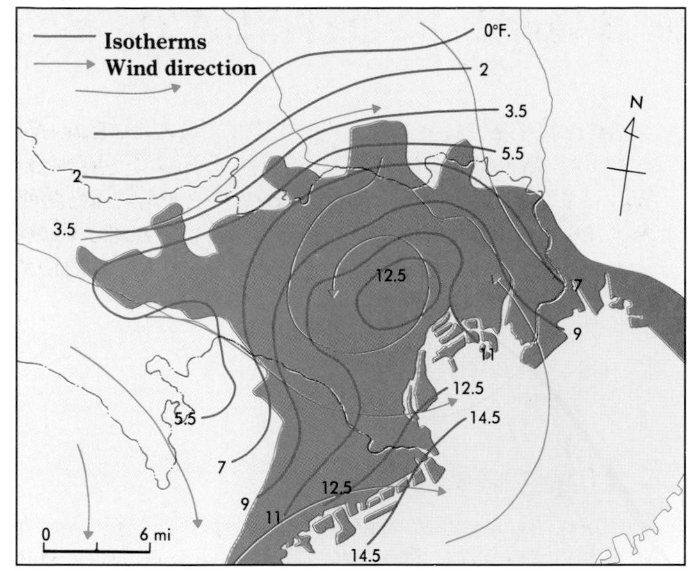


Red lines called isotherms link areas of equal temperature in and around Tokyo, Japan. The temperature is highest near the city's center. The blue lines represent low-altitude winds flowing into the city from the suburbs.

## A heat island is born

Five factors conspire to create a heat island: 1) The heat generated in buildings radiates into the atmosphere. 2) Cars and factories produce carbon dioxide and other gases, inducing a greenhouse effect over the city. 3) Tall buildings inhibit the circulation of warm air and cool air. 4) Building materials such as concrete, asphalt, and steel store heat by day and emit it by night. 5) Sewer systems hinder the evaporation of water, which would help cool the city.

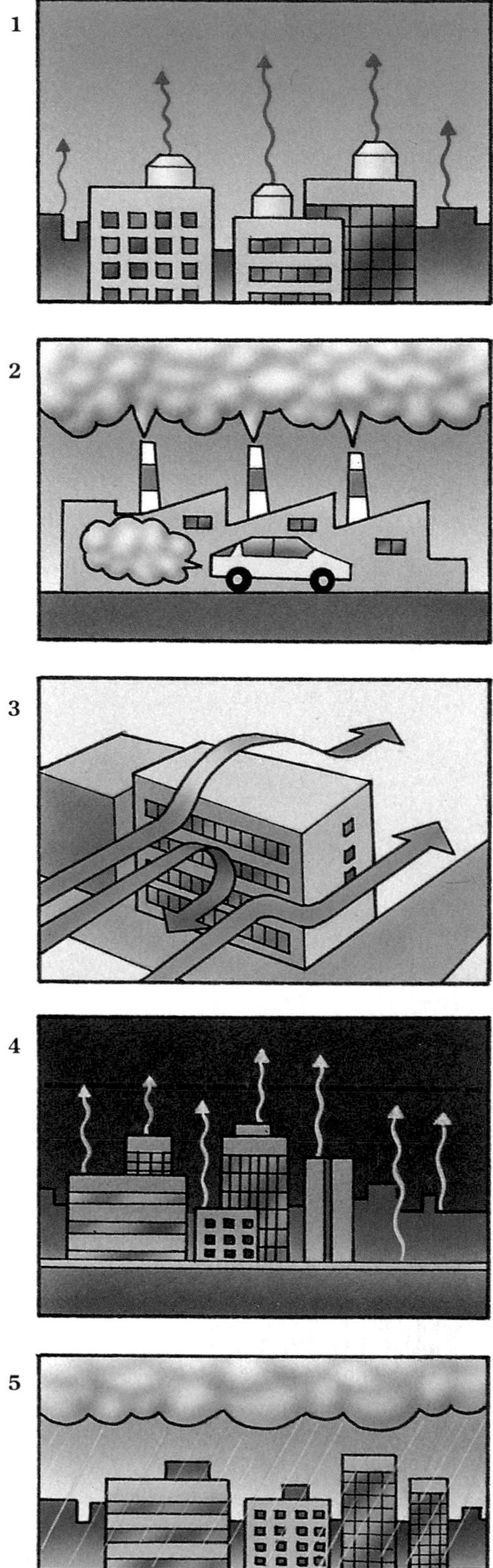

# What Is El Niño?

Every 2 to 10 years, the weather throughout the southern Pacific Ocean shifts wildly. The normally rain-drenched Far East becomes dry, while the arid western coast of South America receives heavy rains, causing *años de abundancia,* or "years of agricultural abundance." Because this phenomenon usually occurs around Christmas, it is called El Niño, the Spanish term for the Christ child.

Scientists cannot fully explain the subtle interactions between ocean and atmosphere that produce El Niño. But they do know that the process involves a weakening of the southeasterly trade winds that normally dictate the area's weather and a corresponding redistribution of warm water throughout the Pacific Ocean.

**Normal weather patterns**

Easterly winds push warm surface water across the Pacific, making its west edge 2° F. warmer and 16 inches higher. To the east, cold water replaces the warm. This creates the pattern below: Warm, moist air rising in the west brings clouds and rain; cold, dry air falling in the east parches the South American coast.

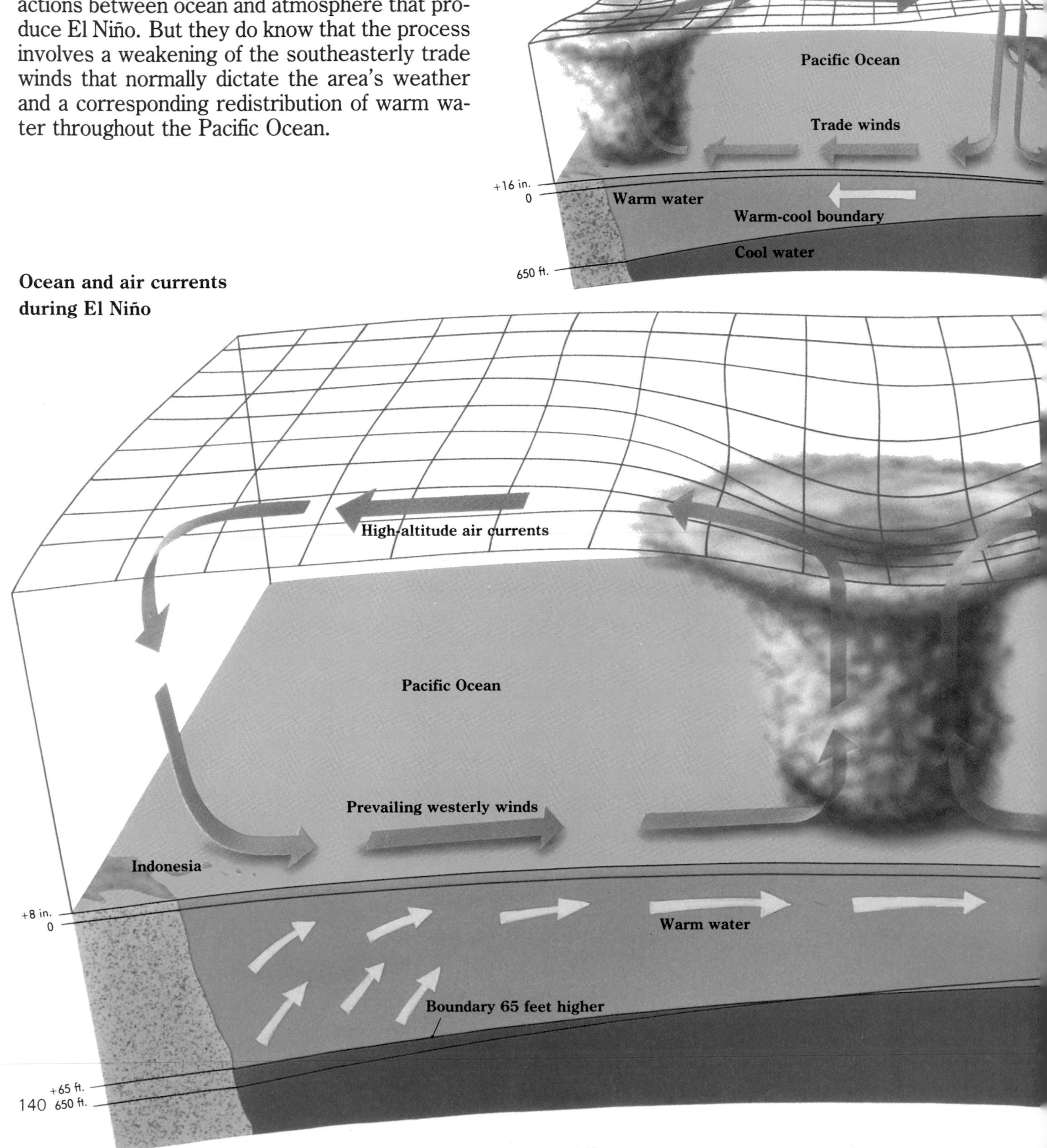


**Ocean and air currents during El Niño**

## A shift in the winds

Through the interplay of wind and water, Earth's southeastern trade winds change from strong to weak and back again in a period of 2 to 10 years. The reason for this shift, known as the southern oscillation, is not known, so the cycles cannot be predicted accurately.

Whereas El Niño represents one extreme of the southern oscillation, the other extreme —called La Niña, Spanish for "the girl"— occurs when air and water currents reinforce each other to produce unusually strong trade winds. As illustrated at right, even more warm water than usual is swept westward across the ocean, bringing torrential rains to the Far East and causing severe droughts in South America.

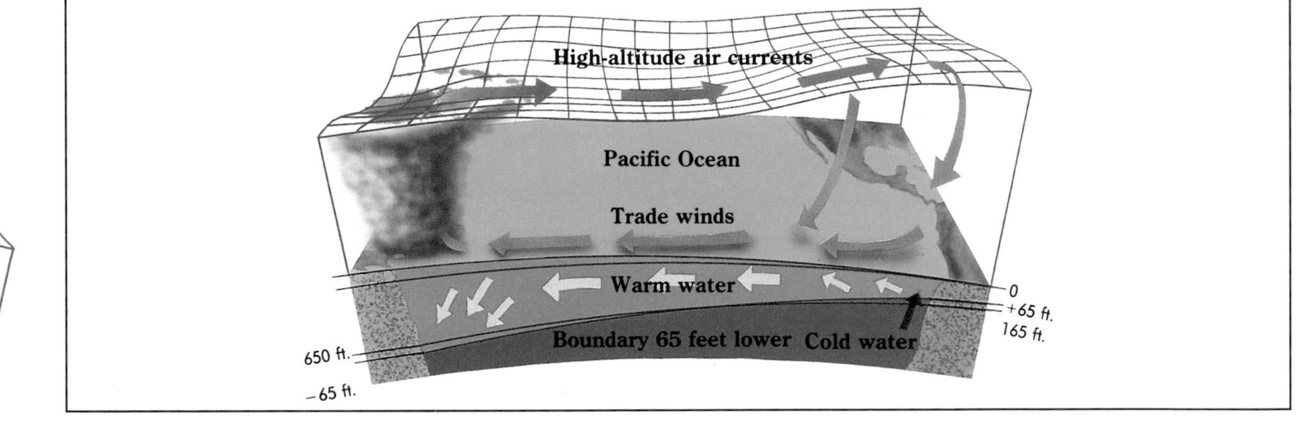


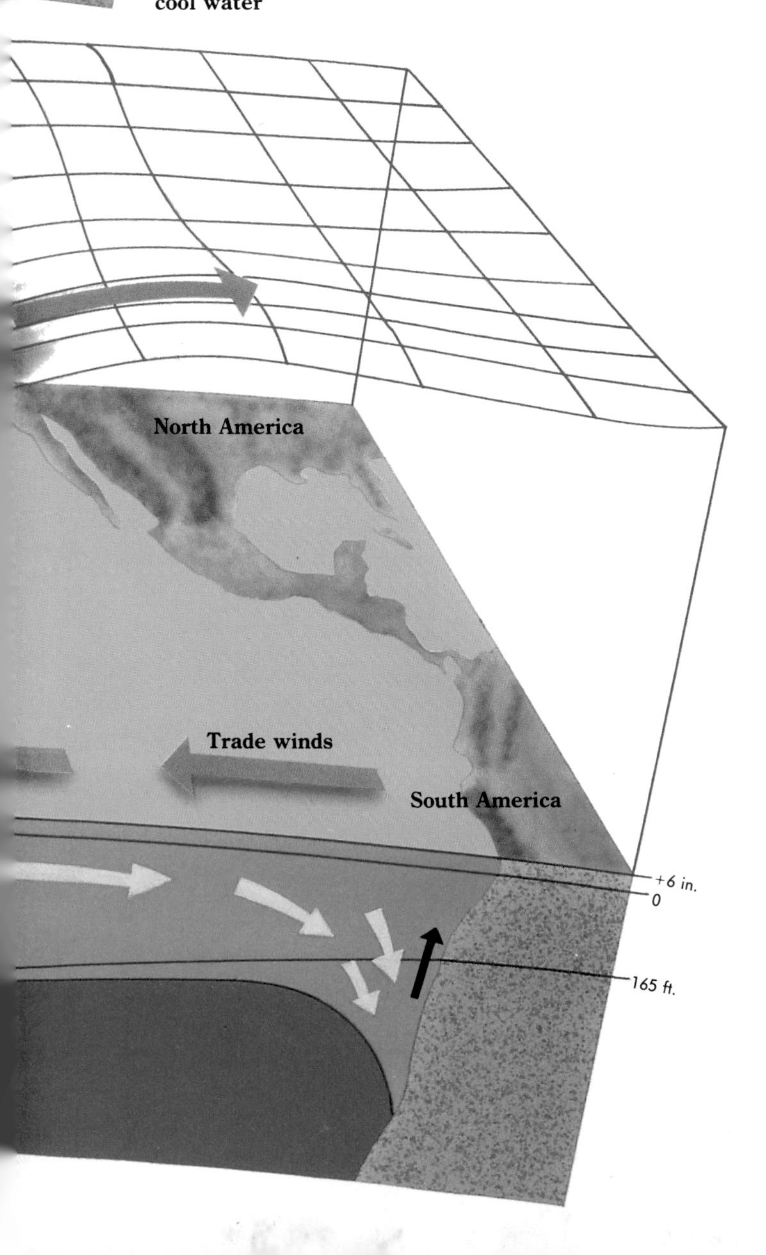


**The onset of El Niño** causes a shift in the weather *(left)*. The prevailing westerly winds ease up, allowing much of the warm water lost to the western side of the Pacific to return east. While the shape of the circulation patterns remains similar, the towering stack of clouds now moves over the eastern Pacific. This robs Asia of its usually heavy rainfall and brings welcome precipitation to western South America.

# How Has Earth's Climate Changed?

Scientists studying geologic clues such as fossils have pieced together a record of Earth's climate for the last 600 million years of its 4.6-billion-year history. During this time, the planet's average temperature has risen and fallen periodically, producing warm thaws and frigid ice ages. Nowadays Earth's average temperature is cooler than its historical average, a reminder that the planet emerged from its last glacial period about 20,000 years ago. Some scientists have suggested that slight variations in Earth's axial tilt *(page 144)* may bring about these temperature fluctuations.

Whatever their cause, climate changes have played a pivotal role in shaping life on Earth. As great amounts of water froze during past ice ages, for example, the sea level dropped, exposing land bridges between continents and allowing many species to spread throughout the world. Anthropologists believe that humans first came to North America during the last ice age, traveling from Asia across the Bering Strait, which was then above sea level.

### Climes of the times

### Continental drift

Some 500 million years ago, geologists believe, the world's continents formed a single landmass. Then, drifting on a global sea of molten rock, they separated, producing today's distribution of land and water.

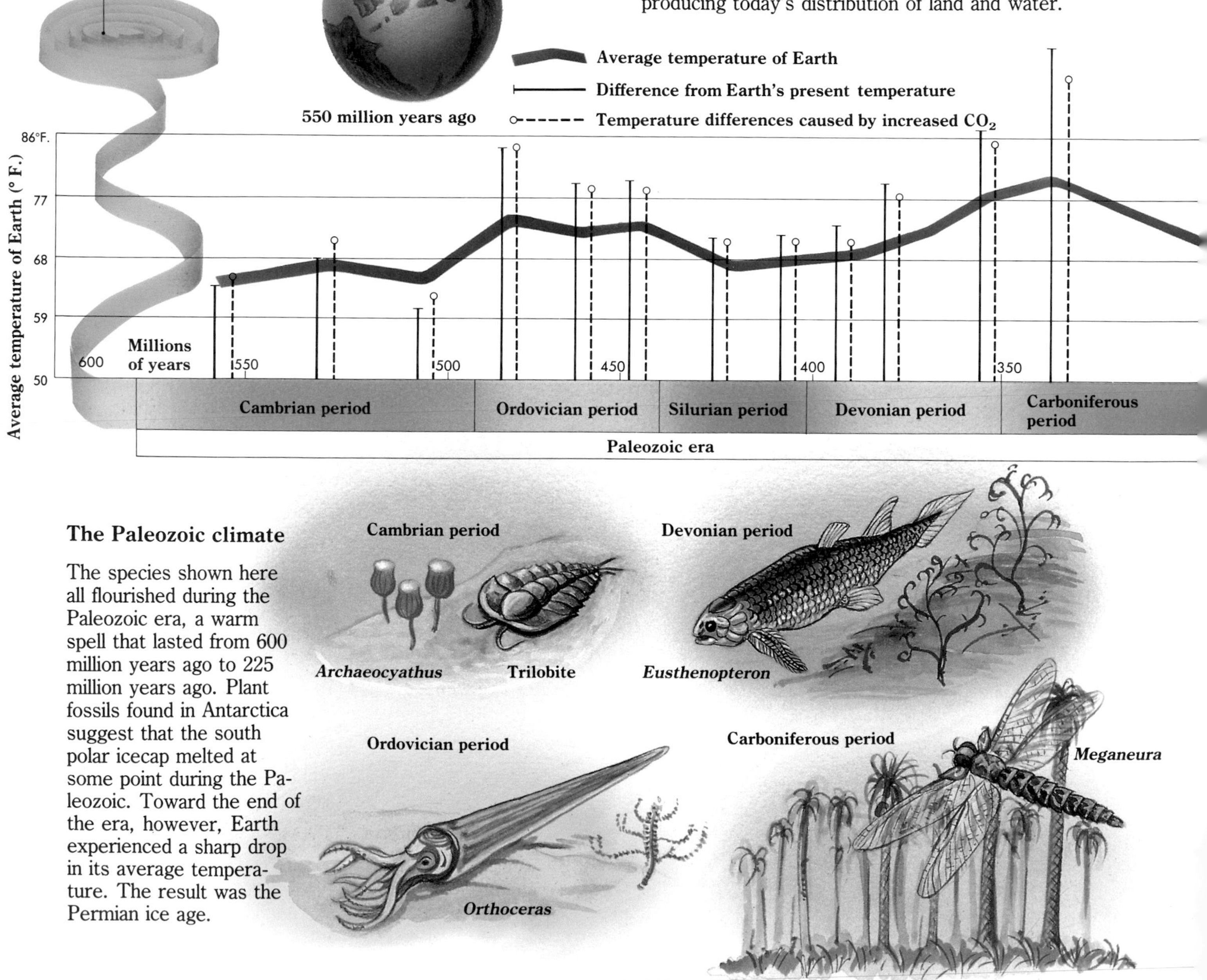


### The Paleozoic climate

The species shown here all flourished during the Paleozoic era, a warm spell that lasted from 600 million years ago to 225 million years ago. Plant fossils found in Antarctica suggest that the south polar icecap melted at some point during the Paleozoic. Toward the end of the era, however, Earth experienced a sharp drop in its average temperature. The result was the Permian ice age.

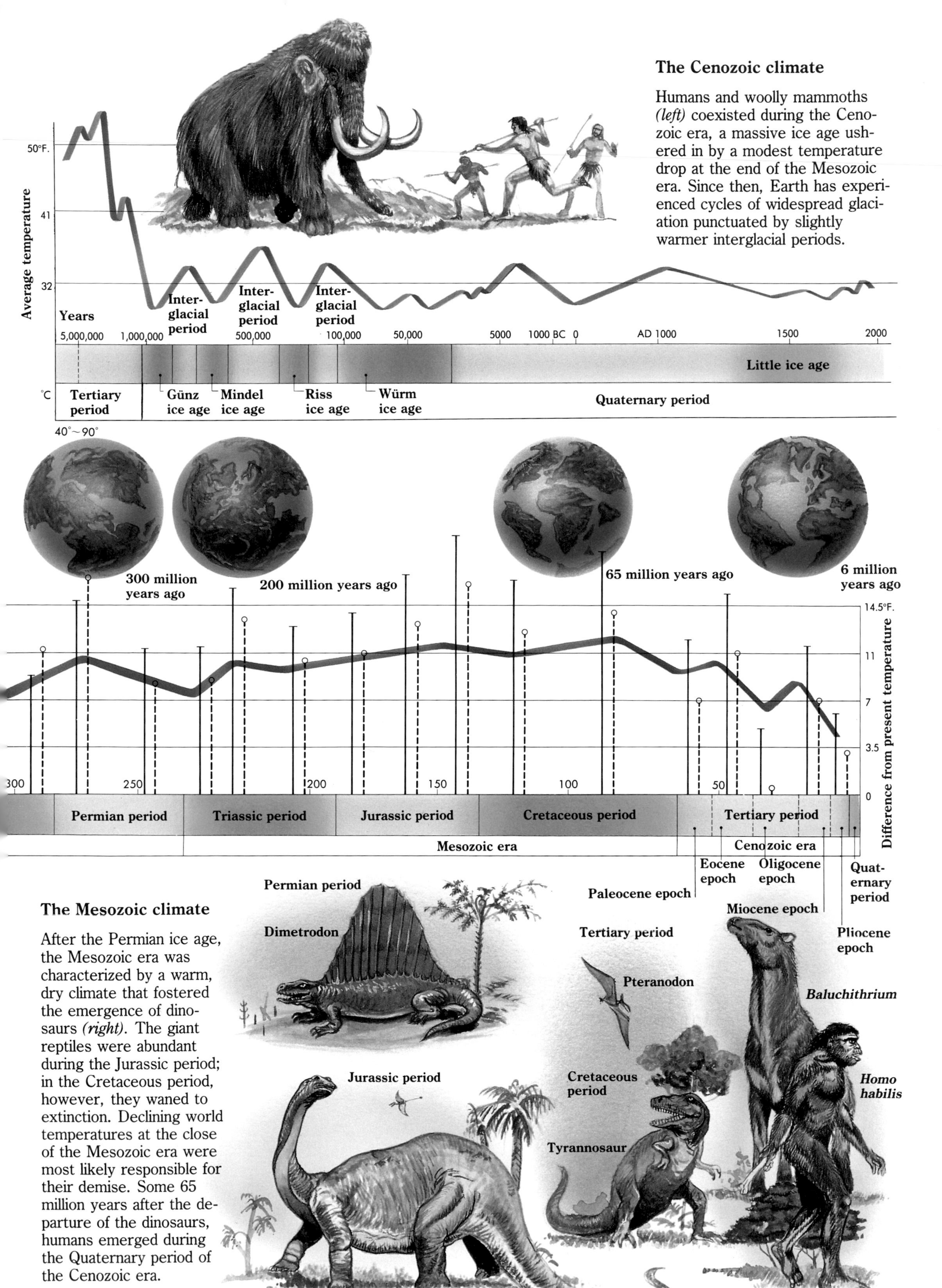


## The Cenozoic climate

Humans and woolly mammoths *(left)* coexisted during the Cenozoic era, a massive ice age ushered in by a modest temperature drop at the end of the Mesozoic era. Since then, Earth has experienced cycles of widespread glaciation punctuated by slightly warmer interglacial periods.

## The Mesozoic climate

After the Permian ice age, the Mesozoic era was characterized by a warm, dry climate that fostered the emergence of dinosaurs *(right)*. The giant reptiles were abundant during the Jurassic period; in the Cretaceous period, however, they waned to extinction. Declining world temperatures at the close of the Mesozoic era were most likely responsible for their demise. Some 65 million years after the departure of the dinosaurs, humans emerged during the Quaternary period of the Cenozoic era.

# What Causes Ice Ages?

Scientists are constantly struggling to pinpoint the reasons for the climate cycles that cause ice ages to come and go. The most famous theory, put forth in 1920 by Yugoslavian mathematician Milutin Milankovitch, argues that Earth's climate is determined by the amount of energy it receives from the sun. According to Milankovitch, this insolation, or incoming solar radiation, is in turn dictated by three astronomical factors.

The first factor is the irregular nature of Earth's orbit around the sun. Over a span of 100,000 years, the planet's orbit changes from a near-perfect circle to a slight oval. As this eccentricity increases, so does the Earth's perihelion, or minimum distance from the sun. The result is less insolation and lower temperatures. The second factor that Milankovitch identified is the tilt of Earth's axis of rotation, which ranges from 21.8° to 24.4° every 20,000 years. When the tilt is greatest, insolation and temperatures both drop. Milankovitch's third factor is the way Earth wobbles on its axis, like a spinning top. This affects the planet's tilt, periodically causing cooler temperatures. When all three factors reinforce one another, said Milankovitch, they plunge the Earth into an ice age.

Although Milankovitch showed that periods of minimum insolation coincided with past ice ages, other scientists have since discovered that astronomical factors alone cannot send temperatures plummeting. Such factors may suffice, however, to trigger a chain of climatic events that could usher in the next ice age. Indeed, computer simulations have shown that the planet could go from its current interglacial period to a full-fledged ice age in just 60,000 years.

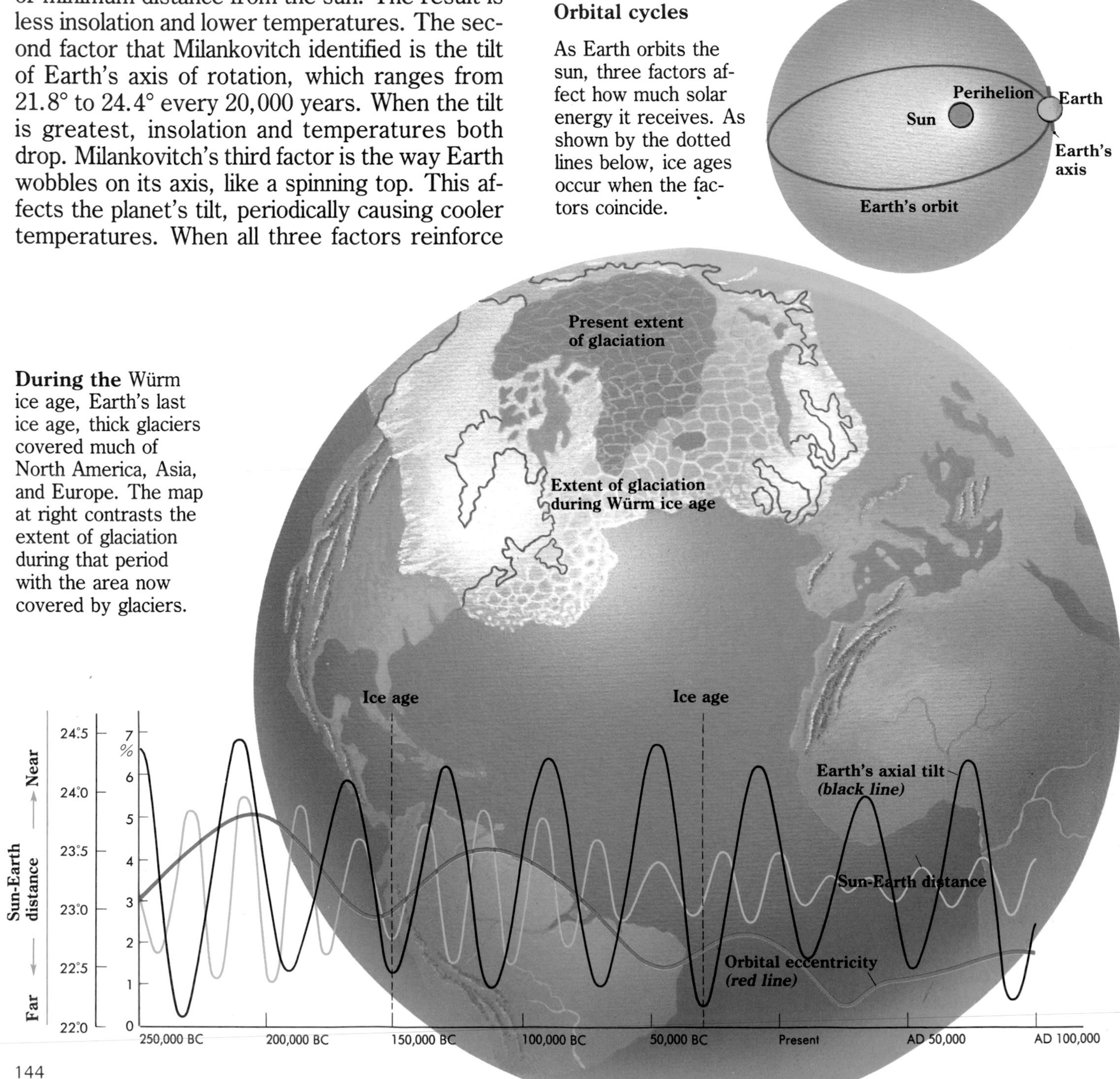


**Orbital cycles**

As Earth orbits the sun, three factors affect how much solar energy it receives. As shown by the dotted lines below, ice ages occur when the factors coincide.

**During the** Würm ice age, Earth's last ice age, thick glaciers covered much of North America, Asia, and Europe. The map at right contrasts the extent of glaciation during that period with the area now covered by glaciers.

# Glossary

**Adiabatic cooling and heating:** The change in temperature of a gas, such as the air, that is caused by a change in pressure, not by the addition or subtraction of heat. In meteorology, the change in pressure is usually caused by parcels of air rising or descending.
**Air mass:** A large body of air with the same temperatures and humidity levels. Air masses often cover thousands of square miles.
**Alto-:** A prefix referring to middle-level clouds.
**Altostratus clouds:** A sheet of thin, grayish or bluish clouds found in the middle levels of the atmosphere.
**Atmosphere:** An envelope of gases surrounding a planet or other celestial body. Earth's atmosphere is divided into five layers, beginning with the **exosphere,** about 400 miles above the Earth, and continuing downward through the **thermosphere, mesosphere, stratosphere,** and **troposphere,** where weather occurs.
**Barometric or pressure gradient force:** A force that results from differences in pressure on either side of a parcel of air, causing the air to move horizontally from high-pressure to low-pressure areas.
**Bow shock:** The area around a planet where the solar wind is first deflected by the planet's magnetosphere.
**Cell:** In meteorology, a part of the atmosphere that acts as a unit.
**Centrifugal force:** An imaginary force that appears to pull an object outward from the center of a circular path.
**Cirrostratus clouds:** Thin sheets of high-level clouds.
**Cirrus clouds:** High-level clouds, often feathery in appearance; sometimes called "mares'-tails."
**Climate:** The long-term weather characteristics of a region.
**Conduction:** The transfer of heat by the collision of molecules. Heat flows from warmer to cooler substances.
**Convection:** The transfer of heat by currents of liquid or air. In meteorology, convection refers to the vertical movement of air.
**Convergence:** The flowing together of air currents.
**Coriolis force:** An apparent force, produced by the rotation of the Earth, that causes the path of a moving object to veer away from a straight line. This force causes winds in the Northern Hemisphere to curve to the right and in the Southern Hemisphere to curve to the left.
**Cosmic rays:** Charged particles, thought to be by-products of violent activities in outer space, that travel close to the speed of light.
**Cumulonimbus clouds:** Low clouds, containing strong updrafts, that tower into the sky and often form anvil-like tops.
**Cumulus clouds:** Detached clouds, formed by rising air currents, that develop in mounded or towering shapes.
**Cyclone:** A low-pressure weather system that can grow into storms such as hurricanes. Driven by the Coriolis force, cyclones blow counterclockwise in the Northern Hemisphere and clockwise in the Southern Hemisphere.
**Diffraction:** The bending of light as it passes around the edge of an object or through narrow slits.
**Divergence:** The splitting of air currents into two or more paths.
**Downdraft:** A column of air that moves downward. An **updraft** is a column of air that moves upward.
**Dropsonde:** A package of instruments, released by airplanes, that takes measurements of the weather.
**Eddy:** Small shifts in air currents.
**Electromagnetic spectrum:** The range of radiation produced by the Sun and other stars. It extends from long-wavelength **radio, infrared,** and **visible** radiation to short-wavelength **ultraviolet** radiation, **x-rays,** and **gamma rays.**
**Equatorial convergence zone:** An area near the equator where the southeast trade winds meet the northeast trade winds.
**Ferrel cell:** A large belt of circulating air that lies between 30° and 60° latitude in the Northern and Southern hemispheres. In this cell, air travels along Earth's surface toward one of the poles, rises at 60° latitude, and flows back toward the equator.
**500-millibar weather map:** A map of the upper air with contour lines showing the height above sea level at which the air pressure is 500 millibars (mb). The standard height of 500 mb is about 18,000 feet, but it can range from 16,000 to 19,000 feet.
**Fossil fuels:** Coal, oil, and natural gas. The name comes from the fact that these fuels are the fossil remains of what was at one time living matter.
**Front:** The boundary between two air masses of different temperatures and humidity.
**Hadley cell:** A large belt of circulating air that lies between the equator and 30° latitude north and south. In this cell, air rises at the equator, moves toward the pole, and sinks at 30° to flow back

toward the equator.
**Ion:** An atom, or group of atoms, that has an electrical charge as a result of being stripped of one or more electrons.
**Ionosphere:** A region of the Earth's atmosphere that contains large numbers of ions. It begins around 40 miles above the Earth's surface.
**Isobar:** A line on a weather map that connects points of equal pressure.
**Isotherm:** A line on a weather map that connects points of equal temperature.
**Jet stream:** A narrow band of high-speed wind in the upper troposphere.
**Latitude:** The distance, in degrees north or south, from the equator. Degrees of latitude begin at zero on the equator and are measured northward to the North Pole at 90° north latitude, and southward to the South Pole at 90° south latitude.
**Lee:** Downwind; or, the side sheltered from the wind.
**Lines of magnetic force:** Imaginary lines that represent an object's magnetic field. These lines run from one pole to the other, parallel to the direction of the magnetic field.
**Magnetic field:** The area around an object that is subject to its magnetism. Earth's magnetic field is similar to that of a bar magnet, with a north and south pole linked by lines of varying magnetic strength and direction.
**Magnetopause:** The boundary between a planet's magnetosheath and its magnetosphere.
**Magnetosheath:** The region in which the solar wind is slowed and deflected around a planet's magnetosphere. The magnetosheath lies between the bow shock and the magnetopause.
**Magnetosphere:** The large region surrounding a planet that is dominated by magnetic field lines.
**Meteorology:** The science that studies the atmosphere and its phenomena. It includes weather forecasting.
**Monsoon:** A seasonal wind, found especially in Asia, that reverses direction between summer and winter and often brings heavy rains.
**Mother-of-pearl clouds:** Thin, shimmering clouds that form in the stratosphere. They are frequently seen at high latitudes, such as in Antarctica, and are also known as **nacreous clouds.**
**Nimbostratus clouds:** Rain clouds thick and extensive enough to block the sun.
**Nimbus clouds:** Rain clouds.
**Noctilucent clouds:** The highest known clouds, observed only at night and at latitudes higher than 50°. Noctilucent clouds appear to form at altitudes of 50 miles and move eastward at high speeds.
**Occluded front:** A type of front that forms when a cold front overtakes a warm front.
**Ozone:** A form of oxygen in which three atoms, instead of the usual two, are bonded together to form a molecule.
**Photochemical reaction:** A chemical reaction that involves light.
**Prevailing westerlies:** The surface winds of the Ferrel cells. Westerlies blow from the west to the east and toward the poles.
**Radiosonde:** A balloon carrying a package of instruments that measures temperature, pressure, and water vapor.
**Reflection:** The return of light from a surface.
**Refraction:** The bending of light as it passes from one substance to another, such as from air to raindrops or glass. Different wavelengths of light bend at different angles; red bends the least and violet the most. This difference causes white light to separate into its different colors.
**Shear:** A cross wind, or a change in the wind in a given direction, such as the increase of wind with height.
**Squall line:** A line of rapidly moving thunderstorms, usually ahead of a cold front.
**Stratus clouds:** Layered, sheetlike clouds formed without vertical movement; also, low-lying uniform clouds.
**System:** Weather phenomena that act as a unit; for example, a high-pressure system.
**Trade winds:** Steady, reliable winds that blow from the east at 30° latitude; also called the **prevailing easterlies.** Trade winds are the surface winds of Hadley cells.
**Weather:** The state of the atmosphere in terms of the six meteorological elements: temperature, humidity, pressure, precipitation, clouds, and wind.
**Wind:** The movement of air relative to the surface of the Earth. Winds are named for the direction from which they blow—northeasterly winds blow out of the northeast, and a land breeze blows from the land toward the sea.
**Windward:** Facing the direction from which the wind is blowing.

# Index

*Numerals in italics indicate an illustration of the subject mentioned.*

# E



# F



# G



# H



# I



# J



# K



# L



# M

**Staff for**
**UNDERSTANDING SCIENCE & NATURE**

*Editorial Directors:* Patricia Daniels, Allan Fallow, Karin Kinney
*Writer:* Mark Galan
*Assistant Editor/Research:* Elizabeth Thompson
*Editorial Assistant:* Louisa Potter
*Production Manager:* Prudence G. Harris
*Senior Copy Coordinator:* Jill Lai Miller
*Production:* Celia Beattie
*Library:* Louise D. Forstall
*Computer Composition:* Deborah G. Tait (Manager), Monika D. Thayer, Janet Barnes Syring, Lillian Daniels

*Special Contributors, Text:* Joseph Alper, Janet Cave, John Clausen, Margery duMond, Patricia Holland, Peter Pocock, Mark Washburn
*Research:* Ruth Williams
*Design/Illustration:* Antonio Alcalá, Caroline Brock, Catherine D. Mason, Nicholas Fasciano, Robert Herndon, Al Kettler, Daniel Rodriguez
*Photography:* Rinhard Eisele (monsoon), NASA (atmosphere), Frederick P. Ostby, National Weather Service (severe storms), Edi Ann Otto (tornado), Lucille Sardegna/Westlight (rainbow), Jim Zuckerman/Westlight (lightning)
*Index:* Barbara L. Klein
*Acknowledgments:* Ken Comba, Colby Hostetler, NOAA

***Consultant:***
Ronald Gird is a meteorologist for the National Weather Service, National Oceanic and Atmospheric Administration (NOAA).

**Library of Congress Cataloging-in-Publication Data**
Weather & climate.
p. cm. — (Understanding science & nature)
Includes index.
Summary: Questions and answers describe the forces involved in the world's weather and climate.
ISBN 0-8094-9683-6 — ISBN 0-8094-9684-4 (lib. bdg.)
1. Weather—Juvenile literature.
2. Climatology—Juvenile literature.
[1. Weather—Miscellanea. 2. Climatology—Miscellanea.
3. Questions and answers.]
I. Time-Life Books. II. Series.
QC981.3.W425 1992
551.5—dc20 91-42442
CIP
AC

**TIME-LIFE for CHILDREN ™**

*Publisher:* Robert H. Smith
*Associate Publisher and Managing Editor:* Neil Kagan
*Editorial Directors:* Jean Burke Crawford, Patricia Daniels, Allan Fallow, Karin Kinney, Sara Mark
*Editorial Coordinator:* Elizabeth Ward
*Director of Marketing:* Margaret Mooney
*Product Manager:* Cassandra Ford
*Assistant Product Manager:* Shelley L. Schimkus
*Director of Finance:* Lisa Peterson
*Financial Analyst:* Patricia Vanderslice
*Administrative Assistant:* Barbara A. Jones
*Special Contributor:* Jacqueline A. Ball

Original English translation by International Editorial Services Inc./ C. E. Berry



Second printing 1993. Printed in U.S.A.
Published simultaneously in Canada.
School and library distribution by Time-Life Education, P.O. Box 85026, Richmond, Virginia 23285-5026.
Time Life Inc. is a wholly owned subsidiary of
THE TIME INC. BOOK COMPANY.
TIME-LIFE is a trademark of Time Warner Inc. U.S.A.
For subscription information, call 1-800-621-7026.